ELECTRONICS SERVICING 500 Q & A FOR PART 2
(Core Studies; Television and Radio Reception; Industrial Equipment)

K. J. Bohlman
T.Eng., F.S.E.R.T., A.M.Inst.E.

Dickson Price Publishers Ltd
Hawthorn House
Bowdell Lane
Brookland, Kent TN29 9RW

Dickson Price Publishers Ltd
Hawthorn House
Bowdell Lane
Brookland
Kent TN29 9RW

First Published 1988
© K. J. Bohlman 1988

British Library Cataloguing in Publication Data

Bohlman, K.J. (Kenneth John)
  Electronics servicing: 500 Q & A for Part 2.
  1. Electronic equipment – Questions &
  answers – For technicians
  I. Title
  621.381'076

ISBN 85380-171-1

All rights reserved. No part of this publication may be reproduced, stored in a retrievel system, or transmitted in any form or by any means electronic, mechanical, photocopy, recording or otherwise, without the prior permission of the publishers.

621.381 BOH
76072

Photoset by
R. H. Services, Welwyn, Hertfordshire
Printed and bound in Great Britain by
Biddles Limited, Guildford and King's Lynn.

cap in series and parallel

ELECTRONICS SERVICING
500 QUESTIONS AND ANSWERS
FOR PART 2

CONTENTS

CORE STUDIES
Semiconductor Diodes   1
Questions
Answers
Transistors and Other Semiconductor Devices   11
Questions
Answers
Voltage and Power Amplifiers   21
Questions
Answers
Power Supplies   37
Questions
Answers
Oscillators   45
Questions
Answers
Logic Circuits and Display Devices   57
Questions
Answers
Differentiating and Integrating Circuits   69
Questions
Answers
L.C.R. Circuits   79
Questions
Answers
Measuring Instruments and C.R.T.   91
Questions
Answers
Transformers and Shielding   105
Questions
Answers
TELEVISION AND RADIO RECEPTION   113
Questions
Answers

INDUSTRIAL EQUIPMENT
Questions
Answers

Other Books of Interest

ELECTRONICS SERVICING VOL 1
ELECTRONICS SERVICING VOL 2
ELECTRONICS SERVICING VOL 3
ELECTRONICS SERVICING 500 MULTIPLE CHOICE QUESTIONS AND ANSWERS FOR PART 1
COLOUR AND MONO TELEVISION VOL 1
COLOUR AND MONO TELEVISION VOL 2
COLOUR AND MONO TELEVISION VOL 3
PRINCIPLES OF DOMESTIC VIDEO RECORDING AND PLAYBACK SYSTEMS
RADIO SERVICING VOL 1
RADIO SERVICING VOL 2
RADIO SERVICING VOL 3
CLOSED CIRCUIT TELEVISION VOL 1
CLOSED CIRCUIT TELEVISION VOL 2

Inspection Copies

Lecturers wishing to examine any of these books should write to the publishers requesting an inspection copy.

Complete Catalogue available on request.

PREFACE

This book contains over 500 questions and answers covering the topic areas of Core Studies, Television and Radio Reception and Industrial Equipment for Part 2 of the City and Guilds 224 Electronic Servicing Course.

The volume caters for the two streams of students (Television and Industrial Electronics) by providing a common Core Studies section and alternative specialist sections. Multiple-choice questions have been used throughout the Core Studies section as this style of assessment is used by the examining body. In the specialist sections, both multiple-choice and short-answer type questions have been included to provide suitable coverage of the essential topics.

It is envisaged that the exercises will be of particular value to the student during the exam revisionary period to test understanding and knowledge of the selected areas of study and to assist in pin-pointing areas of weakness. The Core Studies questions may also be used in class as an assessment aid at the end of each section of learning objectives.

CORE STUDIES

SEMICONDUCTOR DIODES

(1) A common semiconductor material used in the manufacture of p-n diodes is:
 (a) Chromium
 (b) Silicon
 (c) Tungsten
 (d) Selenium.
(2) An example of an **intrinsic** material is:
 (a) Pure germanium
 (b) N-type germanium
 (c) I-type germanium
 (d) Doped germanium.
(3)

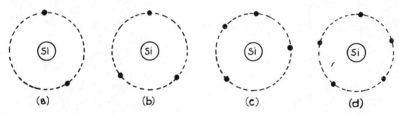

Fig. 1

Which diagram in Fig. 1 shows the correct number of **valency** electrons for an isolated Si atom?
(4) A semiconductor material may typically have a **resistivity** of about:
 (a) 10^{16} ohm-metre
 (b) 10^8 ohm-metre
 (c) 1 ohm-metre
 (d) 10^{-8} ohm-metre.
(5) The number of **valency** electrons associated with an N-type impurity atom are:
 (a) 5
 (b) 4
 (c) 3
 (d) 1.

(6) The number of **valency** electrons associated with a P-type impurity atom are:
  (a) 4
  (b) 1
  (c) 5
  (d) 3.
(7) The **majority** carriers in N-type material are:
  (a) Negative ions
  (b) Positive ions
  (c) Holes
  (d) Electrons.
(8) The addition of a controlled amount of impurity to a pure semiconductor material causes:
  (a) An increase in conductivity
  (b) An increase in thermal sensitivity
  (c) An increase in resistivity
  (d) A decrease in conductivity.
(9)

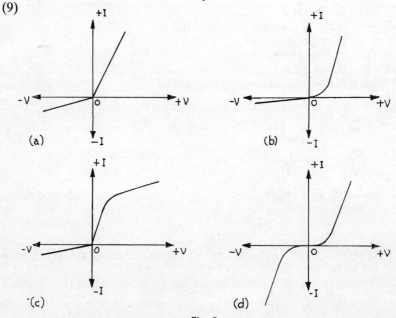

Fig. 2

Which of the diagrams in Fig. 2 shows the correct **voltage-current** characteristic for a p-n diode?

(10) The forward voltage drop of a **silicon** rectifier diode is about:
  (a) 0.2V
  (b) 0.02V
  (c) 700mV
  (d) 5mV.

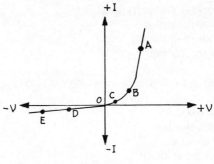

Fig. 3

(11) If measurement of the **slope resistance** is made at each of the lettered points in Fig. 3, which will result in the smallest value being obtained?
(12) The reverse voltage rating for a silicon rectifying diode would be typically:
    (a)   0.8V
    (b)   5V
    (c)   25V
    (d)   1kV.
(13) The forward voltage drop for a **germanium** rectifying diode is about:
    (a)   200mV
    (b)   800mV
    (c)   2V
    (d)   8mV.
(14) Which diagram in Fig. 4 shows the correct **voltage-current** characteristic for a Zener diode?

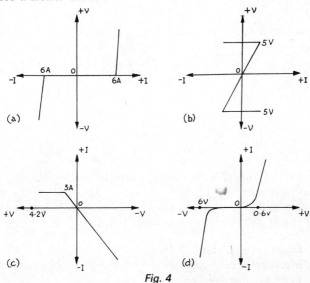

Fig. 4

4 ELECTRONICS SERVICING – QUESTIONS FOR PART 2

(15) A Zener diode is normally used for:
  (a) Power rectification
  (b) Signal demodulation
  (c) Electronic tuning
  (d) Voltage stabilisation.

(16) The variation of **depletion layer** width with reverse voltage is a feature exploited in:
  (a) Rectifier diodes
  (b) Varactor diodes
  (c) Zener diodes
  (d) Point-contact diodes.

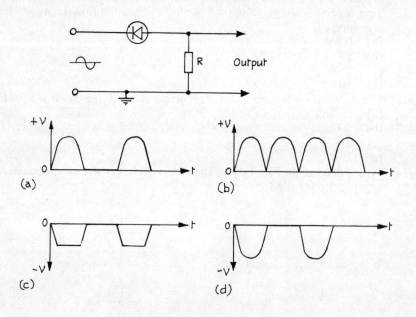

Fig. 5

(17) Which is the correct output voltage waveform diagram for the rectifier circuit shown in Fig. 5?

(18) The maximum voltage that a p-n diode rectifier will withstand in the reverse direction in known as the:
  (a) Zener voltage
  (b) Peak-to-peak value
  (c) Mean reverse voltage
  (d) Peak inverse voltage.

(19) A capacitor is often fitted across a p-n diode rectifier to:
(a) Block the d.c. component
(b) Remove voltage transients
(c) Increase the rectifier capacitance
(d) By-pass the rectifier at mains frequency.

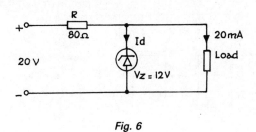

Fig. 6

(20) With reference to Fig. 6, the current I_d flowing in the diode will be:
(a) 80mA
(b) 2mA
(c) 40mA
(d) 230mA.

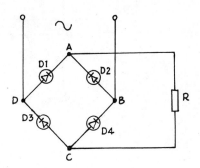

Fig. 7

(21) With reference to Fig. 7 when B is positive with respect to D, the diodes that will be conducting are:
(a) D1 and D4
(b) D2 and D3
(c) D3 and D4
(d) D1 and D2.

6 ELECTRONICS SERVICING – QUESTIONS FOR PART 2

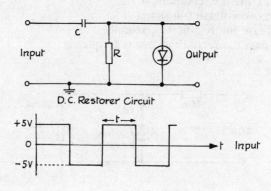

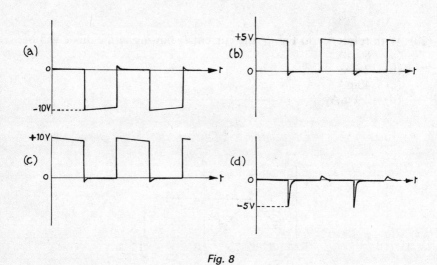

Fig. 8

(22) With reference to Fig. 8, which is the correct output waveform for the d.c. restorer circuit shown?
(23) With reference to Fig. 8, the time-constant of the circuit would be approximately:
 (a) t/5
 (b) t×10
 (c) t/10
 (d) t/100.

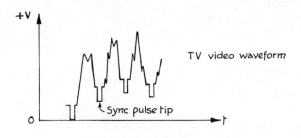

Fig. 9

(24) To ensure that all of the sync. pulse tips of the waveform shown in Fig. 9 are set at the same voltage level, the type of electronic circuit used would be a:
(a) D.C. clipper
(b) D.C. slicer
(c) D.C. restorer
(d) D.C. clamp.

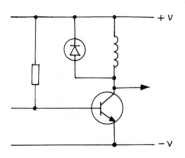

Fig. 10

(25) With reference to Fig. 10, the diode is used to:
(a) Prevent over-voltage between collector and emitter as the transistor cuts-off
(b) Prevent magnetic saturation of the inductor
(c) Prevent the transistor from bottoming
(d) Prevent over-voltage between collector and emitter when the transistor conducts.

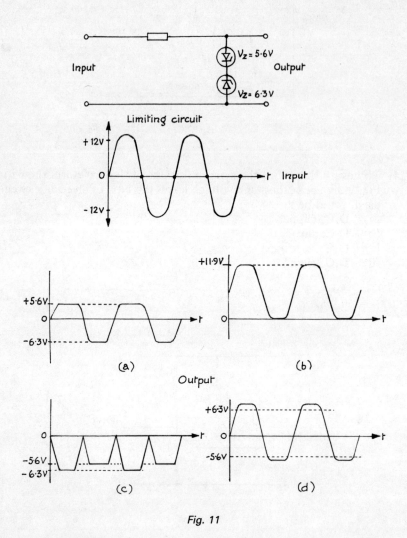

Fig. 11

(26) Which is the correct output waveform for the limiting circuit of Fig. 11?
(27) The ripple frequency at the output of a full-wave rectifier fed from the UK mains supply will be:
 (a) 60Hz
 (b) 100Hz
 (c) 50Hz
 (d) 25Hz.

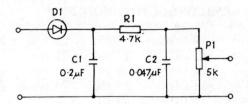

Fig. 12 A.M. demodulator circuit (transistor receiver).

(28) With reference Fig. 12, the output from P1 slider will consist of:
    (a) Modulation component plus d.c. component
    (b) Modulation component only
    (c) Modulation component plus i.f. component
    (d) Modulation component plus d.c. component plus i.f. component.

(29) With reference Fig. 12, the d.c. load for the demodulator is formed by:
    (a) C1 and C2
    (b) C2 only
    (c) R1 plus P1
    (d) R1 only.

(30) With reference Fig. 12, the frequency of the i.f. ripple across C1 will probably be:
    (a) 10.7MHz
    (b) 100kHz
    (c) 5kHz
    (d) 0.47MHz.

ANSWERS ON SEMICONDUCTOR DIODES

1 (b)	6 (d)	11 Point A	16 (b)	21 (b)	26 (d)
2 (a)	7 (d)	12 (d)	17 (d)	22 (a)	27 (b)
3 (c)	8 (a)	13 (a)	18 (d)	23 (b)	28 (a)
4 (c)	9 (b)	14 (d)	19 (b)	24 (d)	29 (c)
5 (a)	10 (c)	15 (d)	20 (a)	25 (a)	30 (d)

CORE STUDIES

TRANSISTORS AND OTHER SEMICONDUCTOR DEVICES

(1) The relationship between base, emitter and collector currents of a bipolar transistor is given by:
 (a) Ie = Ic − Ib
 (b) Ic = Ie + Ib
 (c) Ib = Ic − Ie
 (d) Ie = Ib + Ic.

(2) The a.c. current gain or a bipolar transistor in common emitter will probably be:
 (a) 150
 (b) 0.99
 (c) 0.9
 (d) 10^6.

(3) In normal use the bias state of the junctions in a bipolar transistor are:
 (a) Base-emitter (forward bias) : Collector-base (reverse bias)
 (b) Base-emitter (reverse bias) : Collector-base (reverse bias)
 (c) Base-emitter (reverse bias) : Collector-base (forward bias)
 (d) Base-emitter (forward bias) : Collector-base (forward bias).

(4) A small signal amplifying transistor operating in **common base** would probably have an input resistance of about:
 (a) 10 k-ohm
 (b) 1 M-ohm
 (c) 70 ohm
 (d) 3 k-ohm.

(5) A small signal amplifying transistor operating in **common emitter** would probably have an input resistance of about:
 (a) 1 k-ohm
 (b) 100 k-ohm
 (c) 1 M-ohm
 (d) 10 ohm.

(6) The heat sink of a transistor is painted matt black:
(a) To increase heat radiation
(b) To absorb heat more efficiently
(c) To reflect incident light
(d) To screen the transistor from interference.

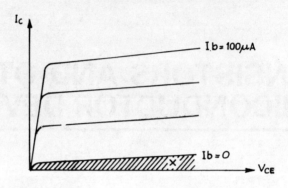

Fig. 13

(7) Refer to Fig. 13. If a transistor is operated in area X then it will be:
(a) Bottomed
(b) Saturated
(c) Cut-off
(d) Overheating.

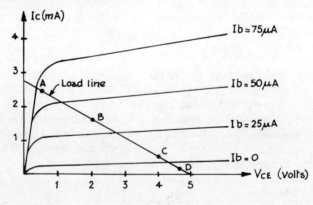

Fig. 14

(8) Refer to Fig. 14. The collector voltage will 'bottom' with a base current of:
(a) 75μA
(b) 50μA
(c) 25μA
(d) Zero.

TRANSISTORS AND OTHER SEMICONDUCTOR DEVICES 13

(9) Refer to Fig. 14. Maximum power will be dissipated in the transistor when the operation is at:
   (a) Point A
   (b) Point B
   (c) Point C
   (d) Point D.

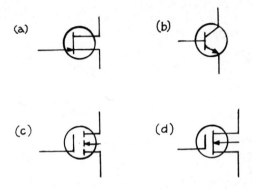

Fig. 15

(10) Refer to Fig. 15. Which diagram shows the correct symbol for an **n-channel depletion mode mosfet**?

(11) The input resistance of a **mosfet** at low frequencies would be about:
   (a) 5 k-ohm
   (b) 200 ohm
   (c) 10^5 ohm
   (d) 10^9 ohm.

(12) A **f.e.t.** is:
   (a) A bipolar device
   (b) A unipolar device
   (c) A current operated device
   (d) A unijunction device.

(13) A mosfet may be supplied with a conductive rubber ring round its leads. The purpose of the ring is to:
   (a) Act as a heat sink
   (b) Prevent damage by an electrostatic charge
   (c) Prevent the leads from bending
   (d) Act as a magnetic screen.

(14) The **common-gate** mode for an **f.e.t.** corresponds to the bipolar transistor
   (a) Common base connection
   (b) Common collector connection
   (c) Common emitter connection
   (d) Emitter follower connection.

14   ELECTRONICS SERVICING – QUESTIONS FOR PART 2

(15) Which of the following would most likely be found in the output stage of a 20W audio amplifier:
   (a)   A VMOS f.e.t.
   (b)   A jugfet
   (c)   A dual gate f.e.t.
   (d)   An n-channel enhancement mosfet.

(16) An advantage of a field effect transistor over a bipolar transistor is:
   (a)   Higher voltage gain
   (b)   Less temperature sensitive
   (c)   Lower input resistance
   (d)   Lower supply voltage requirement.

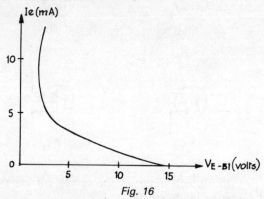

Fig. 16

(17) Refer to Fig. 16. The characteristic shown is typical for:
   (a)   A f.e.t.
   (b)   A unijunction transistor
   (c)   A silicon controlled rectifier
   (d)   A Triac.

(18) A typical application for a U.J.T. would be:
   (a)   Full-wave rectification
   (b)   Half-wave rectification
   (c)   D.C. Stabilisation
   (d)   Trigger device for an S.C.R.

Fig. 17

(19) The circuit symbol shown in Fig. 17 is that for:
   (a)   A U.J.T.
   (b)   A Diac
   (c)   A Triac
   (d)   An S.C.S.

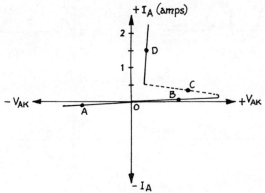

Fig. 18 SCR characteristic.

(20) Refer to Fig. 18. The S.C.R. will be fully 'on' at:
 (a) Point A
 (b) Point B
 (c) Point C
 (d) Point D.
(21) Refer to Fig. 18. If the device is fully 'on', it will revert to the 'off' state when the current through it is reduced to just below:
 (a) 2A
 (b) 1.5A
 (c) 0.5A
 (d) 0.25A.

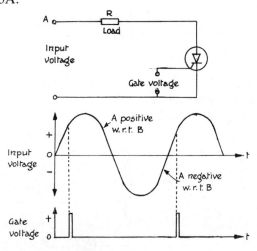

Fig. 19

(22) Refer to Fig. 19 which shows the use of an S.C.R. to control current in a load. Which of the diagrams in Fig. 20 shows the correct current waveshape?

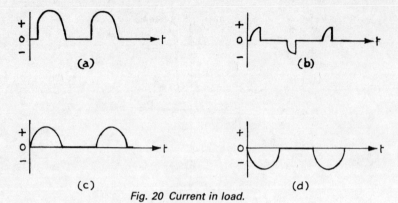

Fig. 20 Current in load.

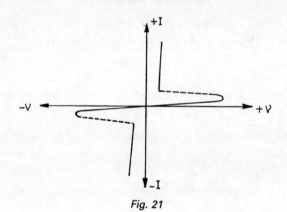

Fig. 21

(23) Refer to Fig. 21. The characteristic shown is typical for:
    (a)   A Triac
    (b)   A Diac
    (c)   A point contact diode
    (d)   A reverse blocking S.C.R.

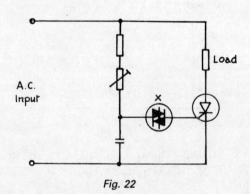

Fig. 22

TRANSISTORS AND OTHER SEMICONDUCTOR DEVICES 17

(24) Refer to Fig. 22. The component marked X is:
 (a) An S.C.R.
 (b) A V.D.R.
 (c) A Diac
 (d) A Triac.

(25) Refer to Fig. 22. Assuming that the a.c. input is sinusoidal then current will flow in the load during:
 (a) Positive half-cycles only
 (b) Negative half-cycles only
 (c) Positive and negative half-cycles
 (d) The transition between following half-cycles.

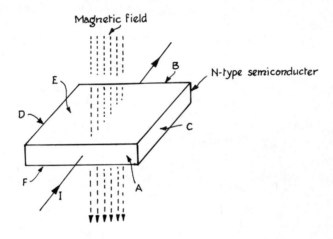

Fig. 23 Hall effect.

(26) Refer to Fig. 23. The Hall voltage will be developed between sides:
 (a) A and B
 (b) C and D
 (c) E and B
 (d) E and D.

(27) A typical application for a Hall Effect device is:
 (a) Over-voltage protection
 (b) Measuring flux in an air-gap
 (c) Storing digital information
 (d) Parity checking.

(28) Microcircuits where complete electronic circuits are formed on a small chip of silicon are known as:
 (a) Printed circuits
 (b) Thick film circuits
 (c) Thin film circuits
 (d) Monolithic integrated circuits.

18 ELECTRONICS SERVICING – QUESTIONS FOR PART 2

(29) One advantage of a pair of output transistors formed by integrated circuit techniques over a pair of discrete transistors is:
    (a) Greater power output
    (b) Smaller heat sink
    (c) Lower output impedance
    (d) Better matching of the pair.

(30) Which of the following components cannot be fabricated in i.c. form:
    (a) 20pF capacitor
    (b) Varicap diode
    (c) 5.6k resistor
    (d) 0.5H inductor.

TRANSISTORS AND OTHER SEMICONDUCTOR DEVICES 19

ANSWERS ON TRANSISTORS AND OTHER SEMICONDUCTOR DEVICES

1 (d)	6 (a)	11 (d)	16 (b)	21 (c)	26 (b)
2 (a)	7 (c)	12 (b)	17 (b)	22 (a)	27 (b)
3 (a)	8 (a)	13 (b)	18 (d)	23 (a)	28 (d)
4 (c)	9 (b)	14 (a)	19 (c)	24 (c)	29 (d)
5 (a)	10 (d)	15 (a)	20 (d)	25 (a)	30 (d)

CORE STUDIES

VOLTAGE AND POWER AMPLIFIERS

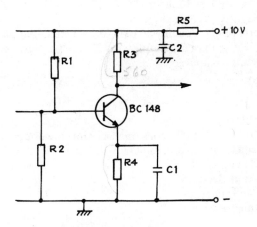

Fig. 24 Voltage amplifier.

(1) Refer to Fig. 24. The mode of connection shown is called:
   (a) Common gate
   (b) Common emitter
   (c) Common collector
   (d) Common base.
(2) Refer to Fig. 24. If the steady voltage drop across the emitter resistor is 0.9V, the voltage at the base terminal with respect to the chassis line will be about:
   (a) − 7.9V
   (b) + 1.5V
   (c) + 0.2V
   (d) − 0.2V.

22 ELECTRONICS SERVICING – QUESTIONS FOR PART 2

(3) Refer to Fig. 24. If the d.c. voltage between collector and chassis is 5.6V and the collector current is 1mA, the value of R3 will be approximately:
  (a) 4.3 k-ohm
  (b) 10 k-ohm
  (c) 5.6 k-ohm
  (d) 560 ohm.

(4) Refer to Fig. 24. The value of R4 is 1.2 k-ohm. If the amplifier is used for audio signal amplification, the value of C1 will probably be about:
  (a) 100nF
  (b) 2μF
  (c) 33μF
  (d) 0.02μF.

(5) Refer to Fig. 24. The d.c. current flowing in R1 will be:
  (a) Zero
  (b) The same as in R2
  (c) Slightly less than in R2
  (d) Slightly greater than in R2.

(6) Refer to Fig. 24. A sine wave of 70mV r.m.s. is applied at the input. If the voltage gain of the amplifier is 20, the peak-to-peak output voltage will be approximately:
  (a) 4V
  (b) 2V
  (c) 1.4V
  (d) 140mV.

(7) Refer to Fig. 24. If C1 goes open-circuit, the effect on the voltage gain of the amplifier to d.c. signals will be:
  (a) No change
  (b) Small decrease
  (c) Large decrease
  (d) Small increase.

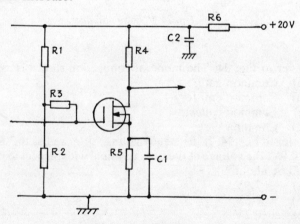

Fig. 25 Voltage amplifier.

VOLTAGE AND POWER AMPLIFIERS

(8) Refer to Fig. 25. The drain load is provided by:
 (a) R1
 (b) R5
 (c) R6
 (d) R4.

(9) Refer to Fig. 25. The purpose of R3 is:
 (a) To prevent negative feedback
 (b) To increase the input impedance of the amplifier
 (c) To reduce the effects of spreads in gm
 (d) To reduce amplifier distortion.

(10) Refer to Fig. 25. If the steady voltage drop across R5 is 2V, the voltage on the gate terminal with respect to the chassis line will probably be about:
 (a) + 0.5V
 (b) + 3.5V
 (c) + 2.7V
 (d) − 3.5V.

(11) Refer to Fig. 25. If the source current is 2mA and the value of R4 is 5.6k-ohm, the d.c. voltage between drain and chassis will be about:
 (a) 9V
 (b) 11V
 (c) 5.6V
 (d) 4.5V.

(12) Refer to Fig. 25. If C1 goes open circuit, the effect on the voltage gain of the amplifier to a.c. signals will be:
 (a) No change
 (b) Decrease
 (c) Doubled
 (d) Trebled.

(13) Refer to Fig. 25. If the gm of the f.e.t. is 3mS and the drain load is 5.6k-ohm the voltage gain of the amplifier will be approximately:
 (a) 1.7
 (b) 17
 (c) 170
 (d) 1700.

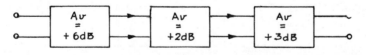

Fig. 26

(14) Refer to Fig. 26. The overall voltage gain of the system will be:
 (a) +11dB
 (b) +1dB
 (c) +36dB
 (d) −36dB.

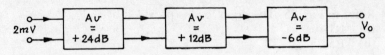

Fig. 27

(15) Refer to Fig. 27. The output voltage (V_0) of the system shown will be approximately:
 (a) 570mV
 (b) 512mV
 (c) 256mV
 (d) 64mV.

(16) The current gain of an amplifier in decibels may be expressed as:
 (a) $10 \log \dfrac{Io}{Ii}$
 (b) $\log \dfrac{Io}{Ii}$
 (c) $20 \log \dfrac{Io}{Ii}$
 (d) $5 \log \dfrac{Io}{Ii}$.

(17) An amplifier provides a power gain of +6dB. If the input power is 100mW, the output power will be approximately:
 (a) 200mW
 (b) 400mW
 (c) 600mW
 (d) 141mW.

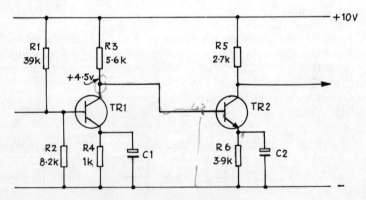

Fig. 28 Two-stage amplifier.

(18) Refer to Fig. 28. The emitter potential of TR2 is probably about:
 (a) + 1.0V
 (b) + 3.9V
 (c) + 1.5V
 (d) + 5.2V.

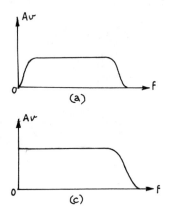

Fig. 29

(19) Refer to Fig. 28. Which of the diagrams in Fig. 29 shows a possible gain-frequency response for the amplifier?
(20) Refer to Fig. 28. If due to, say, a temperature rise the collector current of TR1 increases, a possible effect would be:
  (a) A rise in TR2 emitter voltage
  (b) A rise in TR2 collector voltage
  (c) An increase in TR2 emitter current
  (d) An increase in TR1 collector voltage.
(21) Refer to Fig. 28. The voltage gains of TR1 and TR2 stages are 20 and 12 respectively. The overall voltage gain of the amplifier in decibels will be approximately:
  (a) + 36dB
  (b) + 48dB
  (c) + 240dB
  (d) + 32dB.
(22) Refer to Fig. 28. TR1 and TR2 stages will most likely be working in:
  (a) Class A
  (b) Class C
  (c) Class B
  (d) Class B–C.

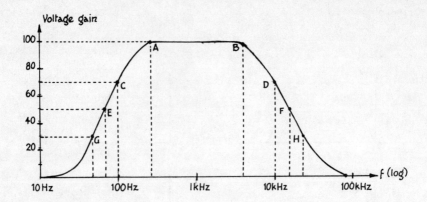

Fig. 30 Response of amplifier.

(23) Refer to Fig. 30. The 3dB bandwidth corresponds to frequencies lying between:
 (a) A and B
 (b) C and D
 (c) E and F
 (d) G and H.
(24) Refer to Fig. 30. The 'half-power' points are:
 (a) A and B
 (b) C and D
 (c) E and F
 (d) A and E.

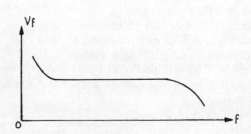

Fig. 31 Negative feedback voltage Vf.

(25) Refer to Fig. 31 which shows how the negative feedback voltage V_f varies with frequency in an amplifier. Which of the diagrams in Fig. 32 shows the correct gain-frequency response?

VOLTAGE AND POWER AMPLIFIERS

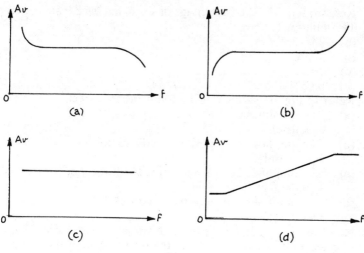

Fig. 32

(26) The voltage gain of an amplifier without feedback is 250. If negative feedback is used and the fraction fed back is $\frac{1}{500}$, the gain with feedback will be approximately:
   (a) 500
   (b) 100
   (c) 279
   (d) 167

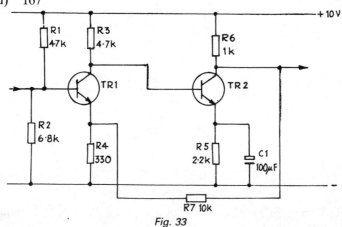

Fig. 33

(27) Refer to Fig. 33. If R7 goes open-circuit the probable consequence will be:
   (a) An increase in TR2 gain
   (b) A decrease in TR2 gain
   (c) An increase in TR1 gain
   (d) A decrease in TR1 gain.

28 ELECTRONICS SERVICING – QUESTIONS FOR PART 2

(28) Refer to Fig. 33. The percentage of signal fed back from the output to TR1 emitter is about:
 (a) 3%
 (b) 100%
 (c) 10%
 (d) 0.03%.

(29) Refer to Fig. 33. Which of the following is correct?
 (a) The amplifier uses negative feedback for a.c. signals only
 (b) The amplifier uses positive feedback for a.c. signals only
 (c) The amplifier uses negative feedback for d.c. signals only
 (d) The amplifier uses negative feedback for a.c. and d.c. signals.

(30) Refer to Fig. 33. If TR2 is replced with a transistor providing a 10% increase in its a.c. current gain, the overall gain of the amplifier will be:
 (a) Increase by 10%
 (b) Increase by less than 10%
 (c) Decrease by more than 10%
 (d) Decrease by less than 10%.

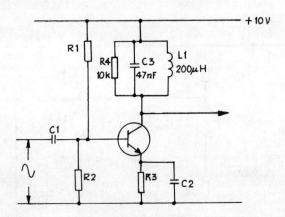

Fig. 34 Voltage amplifier.

(31) Refer to Fig. 34. Maximum voltage gain will occur at a frequency of about:
 (a) 52kHz
 (b) 104MHz
 (c) 16kHz
 (d) 470kHz.

(32) Refer to Fig. 34. Which of the diagrams in Fig. 35 shows the correct gain-frequency response?

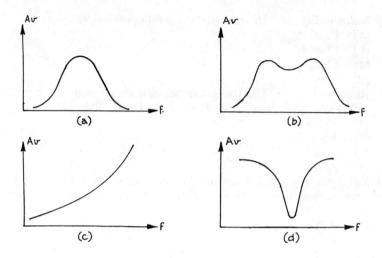

Fig. 35

(33) Refer to Fig. 34. The purpose of R4 is to:
 (a) Increase the amplifier gain
 (b) Increase the amplifier bandwidth
 (c) Increase the Q of the tuned circuit
 (d) Supply the correct collector voltage for the transistor.

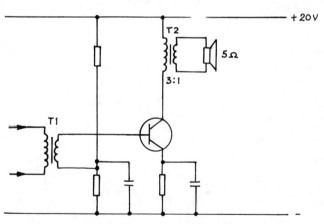

Fig. 36 Power amplifier.

(34) Refer to Fig. 36. The optimum load for the output transistor will be about:
 (a) 5 ohm
 (b) 15 ohm
 (c) 45 ohm
 (d) 1k-ohm.

(35) Refer to Fig. 36. The transistor will be operating in:
 (a) Class A
 (b) Class B
 (c) Class C
 (d) Class D.
(36) Refer to Fig. 36. The transistor passes a steady collector current of 1.5A. Its steady base current will probably be about:
 (a) 15mA
 (b) 1mA
 (c) 200µA
 (d) 20µA.
(37) Refer to Fig. 36. When delivering 3W of audio power to the loudspeaker, the r.m.s. voltage across T1 primary will be about:
 (a) 11.6V
 (b) 9.3V
 (c) 45V
 (d) 1.3V.
(38) Refer to Fig. 36. If the loudspeaker is replaced with one of 10 ohm impedance, the probable effect will be:
 (a) No output
 (b) Increase in distortion and reduction in l.f.
 (c) Decrease in distortion and increase in h.f.
 (d) Excessive transistor dissipation.

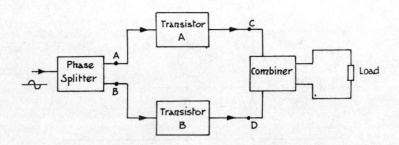

Fig. 37 Class B push-pull block diagram.

(39) Refer to Fig. 37. Which of the diagrams in Fig. 38 shows the correct waveforms for points A and B?

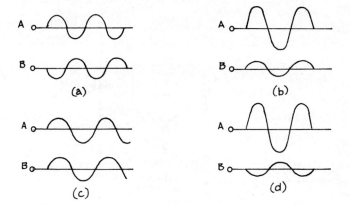

Fig. 38

(40) Refer to Fig. 37. Which of the diagrams in Fig. 39 shows the correct waveforms for points C and D?

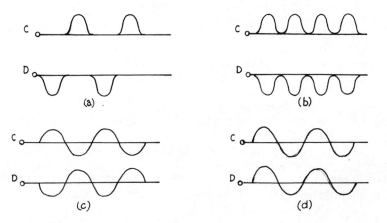

Fig. 39

32 ELECTRONICS SERVICING – QUESTIONS FOR PART 2

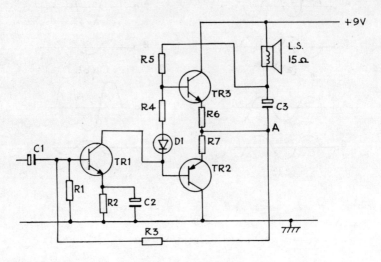

Fig. 40 Class B output stage.

(41) Refer to Fig. 40. With no signal applied, the voltage across C2 will be approximately:
 (a) 8.9V
 (b) 4.5V
 (c) 0.7V
 (d) 1.5V.

(42) Refer to Fig. 40. The purpose of R4 is to:
 (a) Provide a small forward bias for TR2 and TR3
 (b) Act as the load for TR1
 (c) To provide n.f.b. in TR3 stage
 (d) To stabilise the operation of the output transistors.

(43) Refer to Fig. 40. The function of D1 is to:
 (a) To provide symmetrical clipping of the drive signal
 (b) To prevent changes in forward bias with variations in temperature and supply
 (c) To allow TR3 to conduct on the negative half-cycles of the applied signal
 (d) To prevent damage to TR1 should the supply polarity be reversed.

(44) Refer to Fig. 40. If the d.c. voltage present at point A increases slightly, one effect will be for:
 (a) TR1 emitter voltage to fall
 (b) TR1 collector current to fall
 (c) TR2 emitter current to rise
 (d) TR1 collector current to rise.

VOLTAGE AND POWER AMPLIFIERS   33

(45) Refer to Fig. 40. If the 15 ohm loudspeaker is replaced by one of 5 ohm the effect will be:
 (a) Higher dissipation in TR2 and TR3
 (b) Lower dissipation in TR2 and TR3
 (c) Less output power
 (d) Less power taken from d.c. supply.
(46) Refer to Fig. 40. Maximum power will be developed in the loudspeaker when the peak signal voltage between point A and chassis is:
 (a) Exactly 9V
 (b) About 4V
 (c) About 8V
 (d) About 2V.
(47) Refer to Fig. 40. D.C. negative feedback is provided by:
 (a) C3
 (b) C2
 (c) R5
 (d) R3.
(48) Refer to Fig. 40. Which of the waveforms in Fig. 41 shows the effect of reducing the forward bias of the output pair?

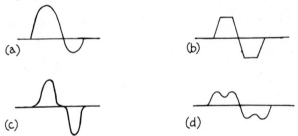

Fig. 41 Output waveform.

(49) Refer to Fig. 40. Due to a fault condition the steady voltage between point A and chassis rises to 6V. Which of the waveforms given in Fig. 42 represents the operation in in these circumstances?

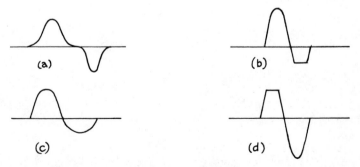

Fig. 42 Output waveform.

(50) An advantage of class B operation over class A operation of a push-pull output stage is:
   (a) Less distortion
   (b) Lower quiescent current
   (c) Anti-phase drives may be dispensed with
   (d) Less n.f.b. need be applied.
(51) An n-p-n transistor is used for 'forward' gain control in a television receiver. If the signal strength increases, the effect on the transistor will be to:
   (a) Increase the collector current
   (b) Decrease the collector current
   (c) Decrease the forward bias
   (d) Make the base potential less positive w.r.t. the emitter.
(52) A p-n-p transistor is used for 'reverse' gain control in a radio receiver. If the signal strength reduces, the effect on the transistor will be to:
   (a) Make the base potential less negative w.r.t. the emitter
   (b) Increase the collector current
   (c) Decrease the emitter current
   (d) Increase the forward bias.

ANSWERS ON VOLTAGE AND POWER AMPLIFIERS

1 (b)	11 (a)	21 (b)	31 (a)	41 (b)	51 (a)
2 (b)	12 (b)	22 (a)	32 (a)	42 (a)	52 (b)
3 (a)	13 (b)	23 (b)	33 (b)	43 (b)	
4 (c)	14 (a)	24 (b)	34 (c)	44 (d)	
5 (d)	15 (d)	25 (b)	35 (a)	45 (a)	
6 (a)	16 (c)	26 (d)	36 (a)	46 (b)	
7 (a)	17 (b)	27 (c)	37 (a)	47 (d)	
8 (d)	18 (b)	28 (a)	38 (b)	48 (c)	
9 (b)	19 (c)	29 (d)	39 (a)	49 (d)	
10 (a)	20 (b)	30 (b)	40 (a)	50 (b)	

CORE STUDIES

POWER SUPPLIES

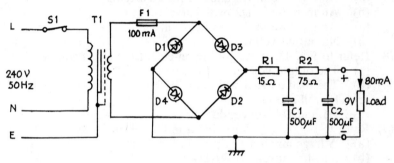

Fig. 43

(1) Refer to Fig. 43. The r.m.s. voltage across T1 secondary will probably be about:
 (a) 200V
 (b) 50V
 (c) 17V
 (d) 12V.
(2) Refer to Fig. 43. The d.c. voltage across C1 will be about:
 (a) 9V
 (b) 9.8V
 (c) 15V
 (d) 35V.
(3) Refer to Fig. 43. The frequency of the ripple voltage across C1 will be:
 (a) 50Hz
 (b) 100Hz
 (c) 150Hz
 (d) 1kHz.
(4) Refer to Fig. 43. The peak inverse voltage across each rectifier will be about:
 (a) 34V
 (b) 10V
 (c) 17V
 (d) 340V.

38   ELECTRONICS SERVICING – QUESTIONS FOR PART 2

(5) Refer to Fig. 43. The reactance of C2 to the ripple voltage will be about:
   (a) 3 ohm
   (b) 75 ohm
   (c) 100 k-ohm
   (d) 2M-ohm.

(6) Refer to Fig. 43. A suitable wattage rating for R2 would be:
   (a) 50mW
   (b) 100mW
   (c) 250mW
   (d) 500mW.

(7) Refer to Fig. 43. If D1 were to go open-circuit there would be:
   (a) An increase in D3 and D4 peak currents
   (b) An increase in D2 peak current
   (c) An increase in the load current
   (d) No output voltage.

(8) Refer to Fig. 43. If R1 goes open-circuit the effect will be:
   (a) The output voltage will fall to about 1V
   (b) The load current will fall to zero
   (c) F1 will rupture
   (d) D1 and D4 will conduct heavily.

(9) Refer to Fig. 43. A short-circuit in C1 will result in:
   (a) A large amount of ripple in the output
   (b) Smaller current in D1-D4
   (c) An increase in load current
   (d) F1 rupturing.

(10) Refer to Fig. 43. An open-circuit C1 will result in:
   (a) No output voltage
   (b) Larger current in D1-D4
   (c) Low output voltage with increase in ripple voltage
   (d) The core of T1 getting excessively hot.

(11) Refer to Fig. 43. The purpose of R1 is:
   (a) Limit the load current at switch-on
   (b) Protect the rectifiers when C1 first charges
   (c) To adjust the output voltage to the correct level
   (d) To remove mains transients.

(12) Refer to Fig. 43. The screen of T1 is connected to earth to:
   (a) Reduce capacitive coupling between primary and secondary
   (b) Reduce magnetic coupling in the transformer
   (c) Assist in keeping the core cool
   (d) Prevent secondary emission.

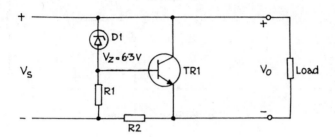

Fig. 44 Shunt stabiliser.

(13) Refer to Fig. 44. If V_s increases the result will be:
 (a) Smaller current in D1
 (b) Larger voltage drop across R2
 (c) Smaller voltage drop across D1
 (d) Smaller current in TR1.
(14) Refer to Fig. 44. If the load current decreases the result will be:
 (a) Less current in TR1
 (b) More current in TR1
 (c) Larger voltage across load
 (d) Smaller voltage drop between TR1 base and emitter.
(15) Refer to Fig. 44. The output voltage of the circuit is probably:
 (a) 15V
 (b) 9V
 (c) 7V
 (d) 6.3V.
(16) Refer to Fig. 44. If the load is short-circuit the outcome will be:
 (a) Excessive dissipation in TR1
 (b) Excessive dissipation in D1
 (c) Power supply fuse ruptures
 (d) R1 goes open-circuit.
(17) Refer to Fig. 44. The output voltage from the stabiliser is equal to:
 (a) Voltage across D1
 (b) Voltage across D1 plus TR1 base-emitter voltage drop
 (c) Voltage across D1 plus voltage across R1
 (d) Voltage across R1 plus TR1 collector-base voltage drop.

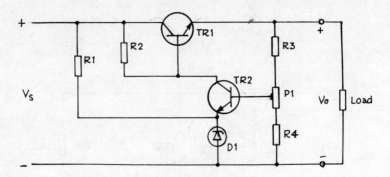

Fig. 45 Series stabiliser.

(18) Refer to Fig. 45. The output voltage from the stabiliser is equal to:
    (a) Voltage across R1 plus voltage across D1
    (b) Voltage across TR1 collector-emitter plus voltage across D1
    (c) V_s minus voltage across D1
    (d) V_s minus voltage across TR1 collector-emitter.

(19) Refer to Fig. 45. A short-circuit across the load will cause:
    (a) TR1 to conduct heavily
    (b) TR2 to conduct heavily
    (c) D1 to conduct heavily
    (d) An increase in V_s.

(20) Refer to Fig. 45. An open-circuit R3 will result in:
    (a) Zero output voltage
    (b) Small reduction in output voltage
    (c) Less current in TR1
    (d) Higher than normal output voltage.

(21) Refer to Fig. 45. An open-circuit R2 will result in:
    (a) Zero output voltage
    (b) High current in TR1
    (c) High current in TR2
    (d) High output voltage.

(22) Refer to Fig. 45. If the load current increases slightly the outcome will be:
    (a) No change in output voltage
    (b) Greater voltage drop across TR1 collector-emitter
    (c) Larger voltage across D1
    (d) Increase in TR2 current.

(23) Refer to Fig. 45. If P1 slider is moved upwards the effect will be:
    (a) Greater current in TR1
    (b) Lower output voltage
    (c) Higher output voltage
    (d) Smaller current in TR2.

POWER SUPPLIES   41

(24) Refer to Fig. 45. A short-circuited D1 will cause:
  (a) Low output voltage
  (b) High output voltage
  (c) Zero output voltage
  (d) Large current in TR1.

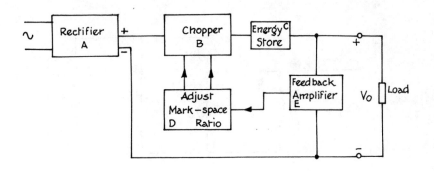

Fig. 46 Switched mode power supply.

(25) Refer to Fig. 46. The device commonly used as the energy store is a:
  (a) Inductor
  (b) Battery
  (c) Zener diode
  (d) R.A.M.
(26) Refer to Fig. 46. The function of block C is:
  (a) To allow the chopper to be disabled on light loads
  (b) To allow continuous power to be fed to the load
  (c) To sample the output voltage
  (d) To act as a voltage doubler.
(27) Refer to Fig. 46. Which of the diagrams in Fig. 47 shows the most probably waveform supplied to the chopper from block D?

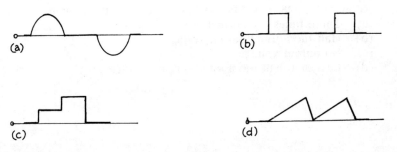

Fig. 47

(28) Refer to Fig. 46. The main advantage of a S.M.P.S. over an ordinary series regulated supply is:
   (a) Less power is required by the load
   (b) Less power is dissipated in the series element
   (c) Smaller value reservoir capacitor required
   (d) Over-voltage protection not required.

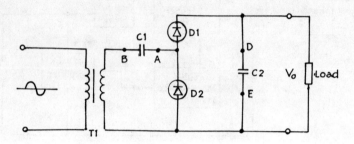

Fig. 48 Cascade voltage doubler.

(29) Refer to Fig. 48. If the r.m.s. voltage across T1 secondary is 100V, the approximate d.c. voltage across C2 will be:
   (a) 140V
   (b) 200V
   (c) 216V
   (d) 282V.
(30) Refer to Fig. 48. The polarities of voltage across C1 and C2 are:
   (a) C1 (A positive to B) and C2 (D postive to E)
   (b) C1 (A negative to B) and C2 (D positive to E)
   (c) C1 (A negative to B) and C2 (D negative to E)
   (d) C1 (A positive to B) and C2 (D negative to E).
(31) Refer to Fig. 48. If D2 becomes short-circuited, the outcome will be:
   (a) Increase in load current
   (b) Half the normal output voltage
   (c) No output voltage
   (d) Greater current in D1.
(32) Refer to Fig. 48. If D2 becomes open-circuit, the outcome will be:
   (a) Three times the normal output voltage
   (b) Half the normal output voltage
   (c) No output voltage
   (d) Less than half the normal output voltage.

ANSWERS ON POWER SUPPLIES

1 (d)	6 (d)	11 (b)	16 (c)	21 (a)	26 (b)	31 (c)
2 (c)	7 (a)	12 (a)	17 (b)	22 (a)	27 (b)	32 (d)
3 (b)	8 (b)	13 (b)	18 (d)	23 (b)	28 (b)	
4 (c)	9 (d)	14 (b)	19 (a)	24 (a)	29 (d)	
5 (a)	10 (c)	15 (c)	20 (d)	25 (a)	30 (a)	

CORE STUDIES

OSCILLATORS

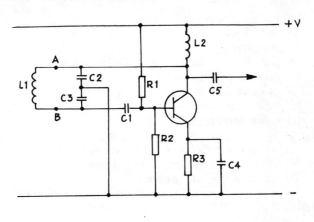

Fig. 49

(1) Refer to Fig. 49. The oscillator circuit shown is called a:
  (a) Hartley oscillator
  (b) Colpitts oscillator
  (c) Reinartz oscillator
  (d) Wein bridge oscillator.
(2) Refer to Fig. 49. The frequency of oscillation is governed by the values of:
  (a) C1, R1
  (b) C1, R2
  (c) L2, C2
  (d) L1, C2, C3.
(3) Refer to Fig. 49. The output waveform from the circuit will be a:
  (a) Sine wave
  (b) Square wave
  (c) Sawtooth wave
  (d) Parabolic wave.

46 ELECTRONICS SERVICING – QUESTIONS FOR PART 2

(4) Refer to Fig. 49. Which of the following statements is correct:
    (a) The oscillator uses positive feedback?
    (b) The oscillator operates in classes 'A'?
    (c) The oscillator uses negative feedback?
    (d) The oscillator can only be used up to frequencies of a few kHz.

(5) Refer to Fig. 49. The purpose of R1, R2 is to:
    (a) Bias the transistor to class 'A'
    (b) Provide a small 'starting' bias
    (c) Provide sliding bias
    (d) Limit the amplitude of oscillations.

(6) Refer to Fig. 49. The phase relationship between the signals at points A and B is:
    (a) In-phase
    (b) 90° out of phase
    (c) 45° out of phase
    (d) In anti-phase.

(7) Refer to Fig. 49. The frequency of oscillation will increase if:
    (a) L1 value decreases
    (b) L2 value increases
    (c) C2 value increases
    (d) C1 value decreases.

(8) Refer to Fig. 49. The function of L2 is:
    (a) To act as the collector load for the oscillator
    (b) To form part of the oscillatory circuit
    (c) To improve the h.f. response
    (d) To prevent the supply impedance from damping the oscillation.

(9) Refer to Fig. 49. If R1 goes open-circuit the consequences will be:
    (a) The frequency of oscillation will be too high
    (b) The frequency of the oscillation will be too low
    (c) The oscillation may not start
    (d) The oscillation will gradually build-up and then die away.

(10) Refer to Fig. 49. 'Sliding bias' is provided by:
    (a) R3, C4
    (b) C1, R1
    (c) C1, R2
    (d) C2, C3.

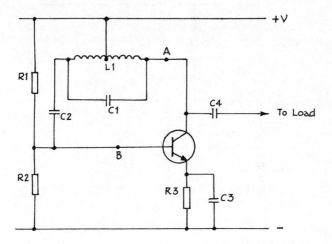

Fig. 50

(11) Refer to Fig. 50. The circuit shown is called a:
 (a) Hartley oscillator
 (b) Colpitts oscillator
 (c) Reinartz oscillator
 (d) Monostable oscillator.
(12) Refer to Fig. 50. For oscillation to occur:
 (a) The loop gain must be equal to or greater than unity
 (b) The tap on L1 must always be at its electrical centre.
 (c) The gain of the transistor must be less than the oscillatory circuit losses
 (d) C1 must be a high Q component.
(13) Refer to Fig. 50. The output waveform from the circuit will be:
 (a) A sawtooth
 (b) A rectangular wave
 (c) A pulse width modulated wave
 (d) A sine wave.
(14) Refer to Fig. 50. The phase relationship between points A and B is:
 (a) 90° out of phase
 (b) Of variable phase
 (c) In-phase
 (d) Anti-phase.
(15) Refer to Fig. 50. A common method of reducing loading effects on the oscillator frequency is:
 (a) Replace C4 with an inductor
 (b) Replace C4 with a zener diode
 (c) Use of buffer stage between oscillator and load
 (d) Reduce the amplitude of the oscillation.

48 ELECTRONICS SERVICING – QUESTIONS FOR PART 2

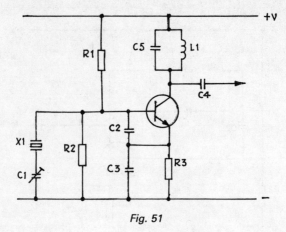

Fig. 51

(16) Refer to Fig. 51. The frequency of oscillation is determined by:
 (a) R1, C2, C3
 (b) X1
 (c) L1, C5
 (d) R1, C2.

(17) Refer to Fig. 51. The operating frequency of the oscillator would probably be:
 (a) 500MHz
 (b) 100MHz
 (c) 1000MHz
 (d) 5MHz.

(18) Refer to Fig. 51. The frequency stability of the oscillator would be typically:
 (a) 1 part in 10^2
 (b) 1 part in 10^3
 (c) 1 part in 10^4
 (d) 1 part in 10^6.

(19) Refer to Fig. 51. The Q of component X1 would be about:
 (a) 10
 (b) 200
 (c) 500
 (d) 20,000.

(20) Refer to Fig. 51. A typical application for the oscillator would be:
 (a) Local oscillator in transistor receiver
 (b) Modulated oscillator
 (c) Master timing
 (d) Bias oscillator in tape recorder.

(21) Refer to Fig. 52. The output waveform from the oscillator will be:
 (a) A sine wave
 (b) A rectangular wave
 (c) A square wave
 (d) A sawtooth.

OSCILLATORS

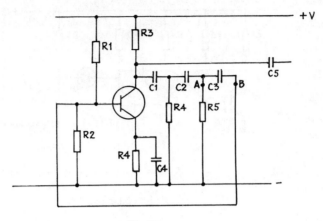

Fig. 52

(22) Refer to Fig. 52. The oscillator circuit is known as:
   (a) An astable oscillator
   (b) A phase-shift oscillator
   (c) A Wein bridge oscillator
   (d) A monostable oscillator.

(23) Refer to Fig. 52. The phase relationship between the signals at points A and B will be:
   (a) Point B leading on point A
   (b) Point B lagging on point A
   (c) In-phase
   (d) In anti-phase.

(24) Refer to Fig. 52. The a.c. current gain of the transistor should be at least:
   (a) 6
   (b) 13
   (c) 23
   (d) 29.

(25) Refer to Fig. 52. A typical operating frequency would be:
   (a) 15Hz
   (b) 1.5MHz
   (c) 0.5MHz
   (d) 5MHz.

(26) The erase head of an audio tape recorder is to be fed from an oscillator. To enable high transfer of oscillatory power to the head the best method would be:
   (a) Use an iron core transformer
   (b) Make the head form part of the oscillatory circuit
   (c) Feed the head via small value capacitors
   (d) Shunt the head with a small value resistor.

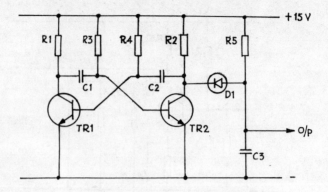

Fig. 53 Simple timebase oscillator.

(27) Refer to Fig. 53. The output waveform from the circuit will probably be:
    (a)   A sawtooth
    (b)   A square wave
    (c)   A sine wave
    (d)   A differentiated wave.

(28) Refer to Fig. 53. The frequency of the output is mainly determined by the values of:
    (a)   C1, R1
    (b)   C1, C2, R3, R4
    (c)   C1, C2, R1, R2
    (d)   C3, R5.

(29) Refer to Fig. 53. There is no output and the voltage at TR2 collector is found to be +15V. A faulty component which could cause these symptoms is:
    (a)   D1 short-circuit
    (b)   R2 open-circuit
    (c)   R3 open-circuit
    (d)   C3 short-circuit.

(30) Refer to Fig. 53. There is zero d.c. voltage across C3 but TR1 and TR2 are found to be operating normally. A faulty component that could produce these symptoms is:
    (a)   D1 short-circuit
    (b)   D1 open-circuit
    (c)   C3 open-circuit
    (d)   R5 open-circuit.

(31) Refer to Fig. 53. When the voltage across C2 rapidly falls:
    (a)   TR1 and TR2 will both be conducting
    (b)   TR1 only will be conducting
    (c)   TR2 and D1 will both be conducting
    (d)   D1 only will be conducting.

OSCILLATORS

(32) Refer to Fig. 53. For correct operation of the circuit:
   (a) TR1 and TR2 will be 'on' for equal periods
   (b) TR1 and TR2 will be 'off' for equal periods
   (c) TR1 will be 'on' for a shorter period than TR2
   (d) TR1 will be 'on' for a longer period than TR2.

(33) Refer to Fig. 53. To synchronise the time base flyback:
   (a) A positive pulse could be applied to D1 anode
   (b) A positive pulse could be applied to TR2 base
   (c) A positive pulse could be applied to TR1 base
   (d) A negative pulse could be applied to TR2 base.

(34) Refer to Fig. 53. If R3 decreases in value the probable outcome would be:
   (a) Reduction in scan period with increase in output frequency
   (b) Reduction in flyback period with increase in output frequency
   (c) Increase in scan period with reduction in output frequency
   (d) Increase in flyback period with increase in output frequency.

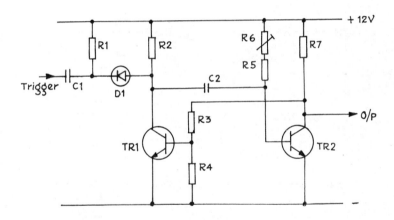

Fig. 54 Monostable.

(35) Refer to Fig. 54. The p.r.f. of the circuit is determined by:
   (a) C2, R2 values
   (b) C2, R5, R6 values
   (c) C1, R1 values
   (d) Trigger p.r.f.

(36) Refer to Fig. 54. Prior to the arrival of a trigger pulse the state of the circuit is:
   (a) TR1 and TR2 both 'on'
   (b) TR1 'on' and TR2 'off'
   (c) TR1 'on' and D1 'on'
   (d) TR1 'off' and TR2 'on'.

52 ELECTRONICS SERVICING – QUESTIONS FOR PART 2

(37) Refer to Fig. 54. The duration of the output pulse is less than normal. A change in component value that would cause this is:
 (a) R4 high in value
 (b) R5 low in value
 (c) C2 high in value
 (d) C1 low in value.

(38) Refer to Fig. 54. There is no output pulse but the trigger input is present. The voltages at TR1 and TR2 collectors are +12V and +0.25V respectively. The most likely component fault is:
 (a) C2 open-circuit
 (b) R5 open-circuit
 (c) R2 open-circuit
 (d) C2 short-circuit.

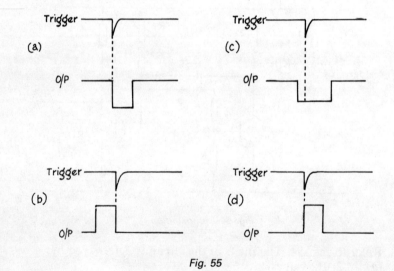

Fig. 55

(39) Refer to Fig. 54. Which of the diagrams in Fig. 55 shows the correct output waveform?

OSCILLATORS

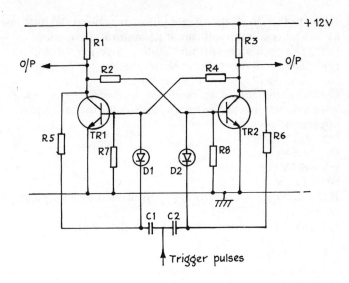

Fig. 56

(40) Refer to Fig. 56. The circuit shown is called:
 (a) A monostable oscillator
 (b) A bistable oscillator
 (c) An astable oscillator
 (d) A tristable oscillator.
(41) Refer to Fig. 56. If the p.r.f. of the trigger input is 5kHz, the output p.r.f. will be:
 (a) 5kHz
 (b) 10kHz
 (c) 1.25kHz
 (d) 2.5kHz.
(42) Refer to Fig. 56. It is usual to include in the circuit two small-value capacitors to increase the switching speed. They would normally be connected across:
 (a) R1 and R3
 (b) R2 and R4
 (c) R5 and R6
 (d) R7 and R8.
(43) Refer to Fig. 56. The output waveforms from TR1 and TR2 collectors are:
 (a) Anti-phase rectangular waves of equal mark-to-space ratios
 (b) Anti-phase rectangular waves of unequal mark-to-space ratios
 (c) In-phase rectangular waves of equal mark-to-space ratios
 (d) In-phase rectangular waves of unequal mark-to-space ratios.

54 ELECTRONICS SERVICING – QUESTIONS FOR PART 2

(44) Refer to Fig. 56. If the trigger pulses are absent the outcome will be:
   (a) The oscillator will run at its 'natural' frequency
   (b) The oscillator will run slightly above its 'natural' frequency
   (c) The oscillator will run slightly below its 'natural' frequency
   (d) The oscillation will fail to start.

(45) Refer to Fig. 56. The oscillator is functioning normally. A d.c. voltmeter connected between TR2 collector and chassis will read approximately:
   (a) 0.25V
   (b) 5.8V
   (c) 8.4V
   (d) 11.2V.

(46) Refer to Fig. 56. Prior to the application of trigger pulses, the initial circuit state is:
   (a) One transistor is 'on' and the other is 'off'
   (b) Both transistors are 'off'
   (c) Both transistors are 'on'
   (d) Both transistors are 'half-on'.

(47) Refer to Fig. 56. The circuit fails to oscillate when trigger is applied. Voltage readings with the trigger disconnected were:
   TR1 collector + 12V
   TR1 base 0V
   TR2 collector + 0.25V
   TR2 base + 0.7V.
   The probable component fault is:
   (a) R2 open-circuit
   (b) R1 open-circuit
   (c) R3 open-circuit
   (d) R4 open-circuit.

(48) Refer to Fig. 56. The function of D1 and D2 is to:
   (a) Ensure that the transistors are cut-off alternately
   (b) Limit the base voltages of the transistors
   (c) Clip the negative trigger pulses
   (d) Limit the rise of collector voltages.

(49) Refer to Fig. 56. When the circuit is functioning normally, TR1 is 'on' for a period of 1ms. The p.r.f. of the output waveform will be:
   (a) 10kHz
   (b) 1kHz
   (c) 500Hz
   (d) 250Hz.

(50) Refer to Fig. 56. With an integrated circuit version of the oscillator, which of the following components would probably not be fabricated in i.c. form:
   (a) C1 and C2
   (b) D1 and D2
   (c) TR1 and TR2
   (d) R1 and R3.

ANSWERS ON OSCILLATORS

1 (b)	11 (a)	21 (a)	31 (c)	41 (d)
2 (d)	12 (a)	22 (b)	32 (d)	42 (b)
3 (a)	13 (d)	23 (a)	33 (b)	43 (a)
4 (a)	14 (d)	24 (d)	34 (a)	44 (d)
5 (b)	15 (c)	25 (a)	35 (d)	45 (b)
6 (d)	16 (b)	26 (b)	36 (d)	46 (a)
7 (a)	17 (d)	27 (a)	37 (b)	47 (d)
8 (d)	18 (d)	28 (b)	38 (a)	48 (a)
9 (c)	19 (d)	29 (c)	39 (d)	49 (c)
10 (a)	20 (c)	30 (d)	40 (b)	50 (a)

CORE STUDIES

LOGIC CIRCUITS AND DISPLAY DEVICES

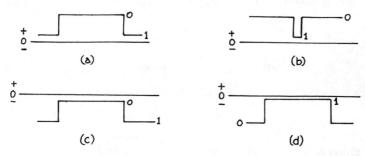

Fig. 57

(1) Refer to Fig. 57. Which of the diagrams shows an example of **positive logic**?

A	B	C	F
0	0	0	1
0	0	1	0
0	1	0	0
0	1	1	0
1	0	0	0
1	0	1	0
1	1	0	0
1	1	1	0

Fig. 58 Truth table.

(2) Refer to Fig. 58. The **Boolean** expression relating to the truth table shown is:
(a) $F = \overline{A+B+C}$
(b) $F = A.B.C$
(c) $F = \overline{A.B.C}$
(d) $F = \overline{A}(B.C)$.

58 ELECTRONICS SERVICING – QUESTIONS FOR PART 2

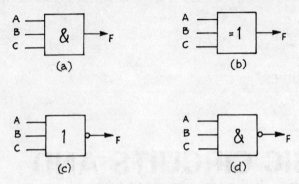

Fig. 59

(3) Refer to Fig. 58. Which of the diagrams in Fig. 59 shows the correct logic symbol relating to the truth table?

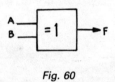

Fig. 60

(4) Refer to Fig. 60. The **Boolean** expression for the gate is:
(a) F = A+B
(b) F = $\overline{A}+\overline{B}$
(c) F = $\overline{A + B}$
(d) F = $\overline{A}B + A\overline{B}$.

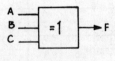

Fig. 61

(5) Refer to Fig. 61. Which of the following input combinations will produce a logic 0 at the output?

	A	B	C
(a)	0	0	1
(b)	1	1	1
(c)	1	1	0
(d)	1	0	1

LOGIC CIRCUITS AND DISPLAY DEVICES

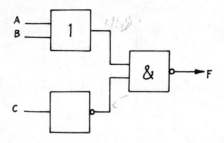

Fig. 62

(6) Refer to Fig. 62. A possible **Boolean** expession for the gate combination is:
 (a) $F = \overline{A+B.C}$
 (b) $F = \overline{A+B}+C$
 (c) $F = \overline{A.B.C}$
 (d) $F = \overline{A.C} + B$.

(7) Refer to Fig. 62. Which of the following input combinations will produce a logic 0 at the output?

	A	B	C
(a)	0	1	1
(b)	1	1	1
(c)	1	0	1
(d)	0	1	0

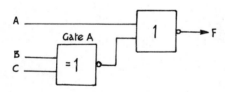

Fig. 63

(8) Refer to Fig. 63. Which of the following input combinations will produce a logic 1 at the output?

	A	B	C
(a)	0	1	0
(b)	1	1	1
(c)	1	1	0
(d)	1	0	0

(9) Refer to Fig. 63. Gate A is:
 (a) A NOR gate
 (b) An exclusive OR gate
 (c) An exclusive NOR gate
 (d) An inequality gate.

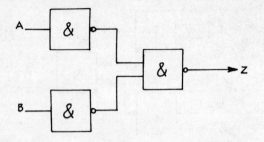

Fig. 64

(10) Refer to Fig. 64. The **Boolean** expression relating to the gate combination is:
(a) Z = A.B
(b) Z = $\overline{A.B}$
(c) Z = A+B
(d) Z = $\overline{A+B}$.

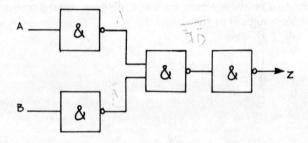

Fig. 65

(11) Refer to Fig. 65. The gate combination shown will carry out the **function** of:
(a) Exclusive OR
(b) NAND
(c) Exclusive NOR
(d) NOR.

(12) Refer to Fig. 65. The **Boolean** expression relating to the gate combination shown is:
(a) Z = $\overline{A+B}$
(b) Z = $\overline{A.B}$
(c) Z = A+B
(d) Z = A+B.

(13) Refer to Fig. 66. The **Boolean** expression relating to the switch diagram is:
(a) F = A.B.C
(b) F = A + B + C
(c) F = A(B + C)
(d) F = A.B + C.

LOGIC CIRCUITS AND DISPLAY DEVICES 61

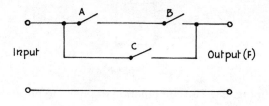

Fig. 66

(14) Which of the following is **true** for an astable oscillator?
    (a)   An astable has two stable states and is triggered
    (b)   An astable has one stable state and is free running
    (c)   An astable has one stable state and is triggered
    (d)   An astable has no stable states and is free running.

(15) Which of the following is **true** for a monostable oscillator?
    (a)   A monostable has one stable state and is triggered
    (b)   A monostable has one stable state and is free running
    (c)   A monostable has two stable states and is triggered
    (d)   A monostable always reverts to any one of two stable states.

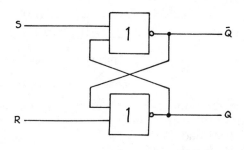

Fig. 67

(16) Refer to Fig. 67. The arrangement will function as:
    (a)   A monostable oscillator
    (b)   A bistable oscillator
    (c)   An astable oscillator
    (d)   A half-adder.

(17) Refer to Fig. 67. An input combination that will produce an indeterminate output is:

	R	S
(a)	0	1
(b)	1	0
(c)	0	0
(d)	1	1.

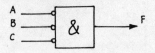

Fig. 68

(18) Refer to Fig. 68. The combination of inputs to produce a **high** output will be:

	A	B	C
(a)	L	L	L
(b)	H	L	H
(c)	H	H	H
(d)	L	H	L

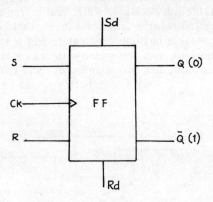

Fig. 69

(19) Refer to Fig. 69. If the logic states at the output are as indicated, the output states will reverse when:
(a) S=1, R=1 and Ck=0
(b) S=1, R=0 and Ck=1
(c) R=1, S=0 and Ck=0
(d) R=0, S=0 and Ck=1.

(20) Refer to Fig. 69. If the logic states at the output are as indicated, the output states will reverse when:
(a) Sd=1, R=0, S=0 and Ck=0
(b) Rd=0, R=1, S=1 and Ck=1
(c) Rd=0, R=1, S=1 and Ck=0
(d) R=1, S=0 and Ck=1.

(21) Refer to Fig. 70. The purpose of gate A is to avoid the indeterminate output of:
(a) S=0 and R=0
(b) S=1 and R=0
(c) S=0 and R=1
(d) S=1 and R=1.

LOGIC CIRCUITS AND DISPLAY DEVICES

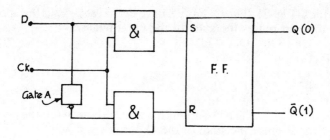

Fig. 70

(22) Refer to Fig. 70. The arrangement shown is often referred to as:
 (a) A synchronous counter
 (b) A latch
 (c) A demultiplexer
 (d) A shift register.

(23) Refer to Fig. 70. If the signal applied to the Ck terminal is missing, the effect will be:
 (a) The output state will reverse on alternate 1s applied to the D input
 (b) The output state will continually oscillate
 (c) The output state will never reverse
 (d) The output state will reverse once and remain in that state.

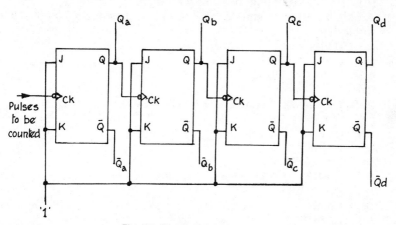

Fig. 71 Ripple through counter.

(24) Refer to Fig. 71. The Q outputs have been reset. After the application of 7 input pulses the state of the Q outputs will be:

	Qa	Qb	Qc	Qd
(a)	0	1	1	1
(b)	1	1	0	1
(c)	1	0	1	1
(d)	1	1	1	0.

64   ELECTRONICS SERVICING – QUESTIONS FOR PART 2

(25) Refer to Fig. 71. The Q outputs have been reset. After the application of 16 input pulses the state of the Q outputs will be:

	Qa	Qb	Qc	Qd
(a)	0	0	0	0
(b)	1	1	1	1
(c)	0	1	1	1
(d)	1	0	0	0.

(26) Refer to Fig. 71. The Q outputs have been reset. After the application of 1 input pulse the state of the $\overline{Q}$ outputs will be:

	$\overline{Qa}$	$\overline{Qb}$	$\overline{Qc}$	$\overline{Qd}$
(a)	0	1	1	1
(b)	1	1	1	0
(c)	1	0	0	1
(d)	0	1	1	0.

(27) Refer to Fig. 71. For every 32 pulses applied to the input there will be:
 (a) 8 pulses at Qa
 (b) 8 pulses at Qb
 (c) 8 pulses at Qc
 (d) 8 pulses at Qd.

(28) Refer to Fig. 71. The Qa output will reverse its state:
 (a) On alternate input pulses only
 (b) Each time the input pulse goes low
 (c) Each time the input pulse goes high
 (d) Each time the input pulse goes high or low.

(29) A suitable logic circuit for use in **switch debouncing** applications would be:
 (a) Half-adder
 (b) Exclusive NOR
 (c) Exclusive OR
 (d) R-S bistable.

(30) Refer to Fig. 72. The decoder inputs ABCD will be in:
 (a) Binary
 (b) BCD
 (c) Decimal
 (d) Hexadecimal.

(31) Refer to Fig. 72. The decoder outputs a-g will be in:
 (a) Decimal
 (b) Binary
 (c) BCD
 (d) Octal.

(32) Refer to Fig. 72. When the display is indicating the numeral seven, the state of the decoder inputs will be:

	A	B	C	D
(a)	0	1	1	1
(b)	0	7	0	0
(c)	1	1	1	0
(d)	0	0	6	1.

LOGIC CIRCUITS AND DISPLAY DEVICES

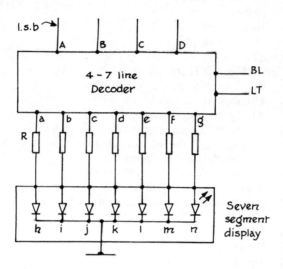

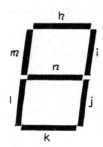

Fig. 72 Seven segment LED display and decoder.

(33) Refer to Fig. 72. When the display is indicating the numeral two, the state of the decoder outputs will be:

	a	b	c	d	e	f	g
(a)	L	L	H	H	L	L	L
(b)	H	H	L	H	H	L	H
(c)	H	H	H	H	L	L	H
(d)	L	L	H	L	L	H	L

(34) Refer to Fig. 72. When the display is indicating the numeral one, the state of the diodes will be:
   (a) Diodes m and l – 'on'
   (b) Diodes i and j – 'on'
   (c) Diode n – 'on'
   (d) Diode k – 'on'.

66   ELECTRONICS SERVICING – QUESTIONS FOR PART 2

(35) Refer to Fig. 72. When any diode is emitting light, the voltage drop across it will be about:
 (a) 10V
 (b) 2mV
 (c) 5V
 (d) 2V.

(36) Refer to Fig. 72. If the LT terminal is taken 'low' the effect will be:
 (a) Decoder outputs a-g will go 'low'
 (b) The display will be blank
 (c) All the segments will light
 (d) There will be no supply to the decoder.

(37) Refer to Fig. 72. If the BL terminal is taken 'low' the effect will be:
 (a) The display will be blank
 (b) Decoder outputs a-g will go 'high'
 (c) All the segments will light
 (d) All of the diodes will be forward biassed.

(38) Refer to Fig. 73. The decoder inputs ABCD will be in:
 (a) BCD
 (b) Binary
 (c) Octal
 (d) Hexadecimal.

(39) Refer to Fig. 73. If the decoder input to any of the exclusive – or gates is 'high', its output will be:
 (a) In anti-phase with the back plate voltage
 (b) 90° out of phase with the back plate voltage
 (c) In phase with the back plate voltage
 (d) 45° out of phase with the back plate voltage.

(40) Refer to Fig. 73. Segments h,i,n,l and k appear opaque when light falls on the display. The state of the decoder outputs will be:

	a	b	c	d	e	f	g
(a)	L	L	H	L	L	H	L
(b)	L	H	H	H	H	L	L
(c)	L	L	L	L	L	L	L
(d)	H	H	L	H	H	L	H

(41) The total current taken from the supply when all of the segments are active in a seven-segment liquid crystal display would be typically:
 (a) 100mA
 (b) 10mA
 (c) 0.2A
 (d) 10µA.

(42) The maximum reverse voltage that may be applied to a L.E.D. is about:
 (a) 50V
 (b) 0.2V
 (c) 5V
 (d) 500V.

LOGIC CIRCUITS AND DISPLAY DEVICES

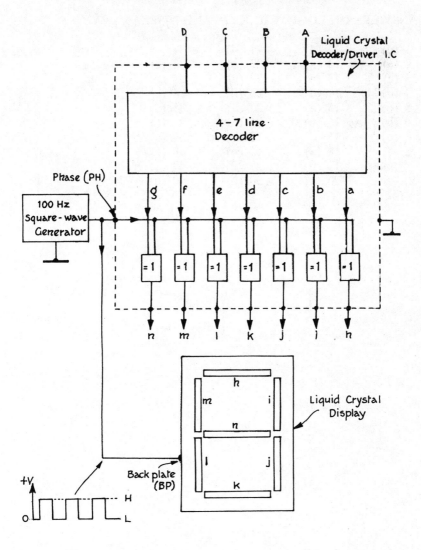

Fig. 73 Seven segment liquid crystal display and decoder.

ANSWERS ON LOGIC CIRCUITS AND DISPLAY DEVICES

1 (d)	11 (d)	21 (d)	31 (b)	41 (d)
2 (a)	12 (a)	22 (b)	32 (c)	42 (c)
3 (c)	13 (d)	23 (c)	33 (b)	
4 (d)	14 (d)	24 (d)	34 (b)	
5 (b)	15 (a)	25 (a)	35 (d)	
6 (b)	16 (b)	26 (a)	36 (c)	
7 (d)	17 (d)	27 (b)	37 (a)	
8 (a)	18 (a)	28 (b)	38 (a)	
9 (c)	19 (b)	29 (d)	39 (a)	
10 (c)	20 (a)	30 (b)	40 (d)	

CORE STUDIES

DIFFERENTIATING AND INTEGRATING CIRCUITS

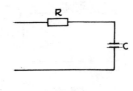

Fig. 74

(1) Refer to Fig. 74. The circuit has a time constant of 10ms. Possible values for R and C are:
 (a) 5 k-ohm and 20µF
 (b) 10 k-ohm and 1nF
 (c) 100 k-ohm and 1nF
 (d) 2 k-ohm and 5µF.

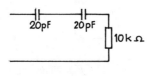

Fig. 75

(2) Refer to Fig. 75. The time constant of the circuit is:
 (a) 100ns
 (b) 400ns
 (c) 400µs
 (d) 100µs.

70 ELECTRONICS SERVICING – QUESTIONS FOR PART 2

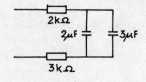

Fig. 76

(3) Refer to Fig. 76. The time constant of the circuit is:
   (a) 250μs
   (b) 4.3ms
   (c) 10ms
   (d) 25ms.

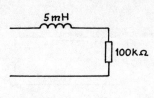

Fig. 77

(4) Refer to Fig. 77. The time constant of the circuit is:
   (a) 500s
   (b) 5s
   (c) 500ns
   (d) 50ns.

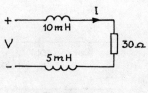

Fig. 78

(5) Refer to Fig. 78. The time taken for the current I to reach 0.63 of its final value will be:
   (a) 450ms
   (b) 4.5ms
   (c) 0.5ms
   (d) 0.315ms.

DIFFERENTIATING AND INTEGRATING CIRCUITS 71

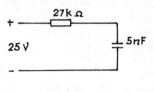

Fig. 79

(6) Refer to Fig. 79. After a period equal to 135µs from connecting the supply, the voltage across the capacitor will be approximately:
 (a) 15.75V
 (b) 12.5V
 (c) 9.25V
 (d) 25V.
(7) Refer to Fig. 79. After a period equal to five times the time-constant from connecting the supply, the voltage across the resistor will be:
 (a) 0V
 (b) 25V
 (c) 9.25V
 (d) 5V.
(8) Refer to Fig. 79. After a period equal to five time the time-constant from connecting the supply, the voltage across the capacitor will be:
 (a) 0V
 (b) 10V
 (c) 19V
 (d) 25V
(9) Refer to Fig. 79. The maximum current that will flow in the circuit is about:
 (a) 0.8mA
 (b) 0.93mA
 (c) 5A
 (d) 80mA.
(10) Refer to Fig. 79. After a period equal to 135µs from connecting the supply, the voltage across the resistor will be about:
 (a) 25V
 (b) 15.75V
 (c) 9.25V
 (d) 0V.

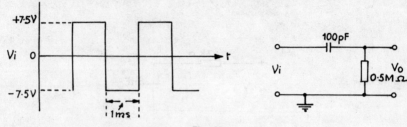

Fig. 80

(11) Refer to Fig. 80. Which of the diagrams given in Fig. 81 shows the correct output waveform from the circuit?

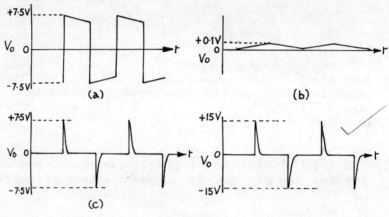

Fig. 81

(12) Refer to Fig. 80. If the positions of the resistor and capacitor are interchanged, which of the diagrams in Fig. 82 will show the correct output waveform?

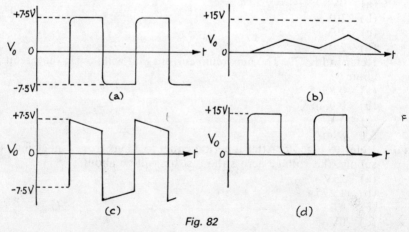

Fig. 82

DIFFERENTIATING AND INTEGRATING CIRCUITS 73

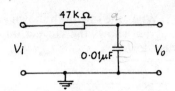

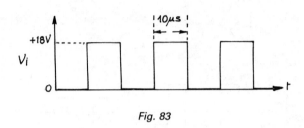

Fig. 83

(13) Refer to Fig. 83. When the square wave shown is applied to the input, which diagram in Fig. 84 shows the correct output waveform after the circuit has settled down?

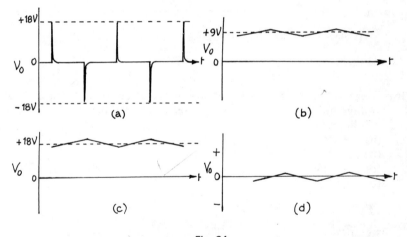

Fig. 84

(14) Refer to Fig. 83. Under the conditions referred to in question 13, the voltage indicated on a d.c. voltmeter connected across the capacitor will be:
(a) +18V
(b) + 9V
(c) 0V
(d) +4.5V.

74 ELECTRONICS SERVICING – QUESTIONS FOR PART 2

(15) Rever to Fig. 83. When the sawtooth shown in Fig. 85 is applied to the input, which diagram shows the correct output waveform after the circuit has settled down?

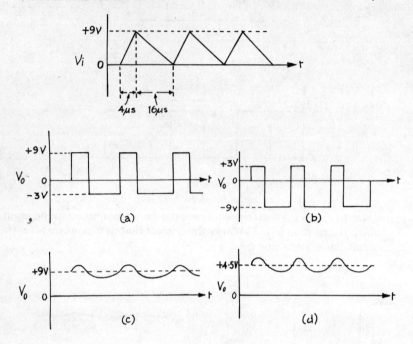

Fig. 85

(16) Refer to Fig. 83. Under the conditions referred to in question 15, the voltage indicated on a d.c. voltmeter connected across the capacitor will be:
 (a) +2V
 (b) −2V
 (c) +4.5V
 (d) +9V.

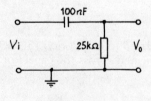 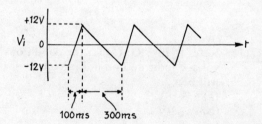

Fig. 86

DIFFERENTIATING AND INTEGRATING CIRCUITS

(17) Refer to Fig. 86. Which of the diagrams given in Fig. 87 shows the correct output waveform?

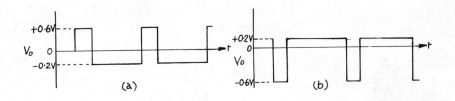

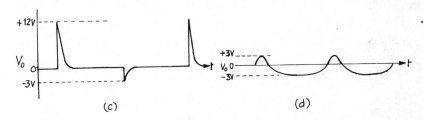

Fig. 87

(18) An **integrator** is:
   (a) A band pass filter
   (b) A low pass filter
   (c) A high pass filter
   (d) A notch filter.

(19) A **differentiator** is:
   (a) A high pass filter
   (b) A band pass filter
   (c) A tuned filter
   (d) A low pass filter.

(20) A repetitive pulse waveform having a pulse duration of 12μs is fed to a CR network. A differentiated output will appear across R if the CR time is:
   (a) 12μs
   (b) 120μs
   (c) 1.2ms
   (d) 1μs.

(21) An integrating network is fed with a square wave. Its output will be:
   (a) A sawtooth wave
   (b) A parabolic wave
   (c) A triangular wave
   (d) A sine wave.

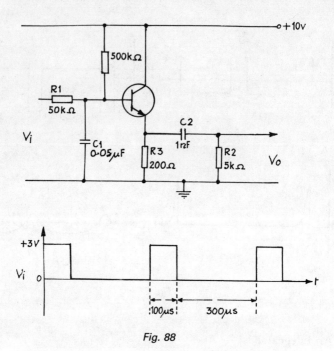

Fig. 88

(22) Refer to Fig. 88. The waveform across R3 will be:
   (a) A large amplitude rectangular wave
   (b) A small amplitude parabolic wave
   (c) A small amplitude sawtooth wave
   (d) Small amplitude positive and negative going spikes.
(23) Refer to Fig. 88. The output waveform (Vo) will be:
   (a) A small amplitude parabola
   (b) A very small amplitude rectangular wave
   (c) A large amplitude sawtooth wave
   (d) A very small amplitude triangular wave.
(24) Refer to Fig. 88. If C1 goes open-circuit, the output waveform (Vo) will be:
   (a) A sawtooth wave
   (b) A rectangular wave
   (c) Positive and negative going spikes
   (d) A parabola.

DIFFERENTIATING AND INTEGRATING CIRCUITS 77

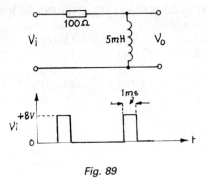

Fig. 89

(25) Refer to Fig. 89. Which of the diagrams in Fig. 90 shows the correct output waveform?

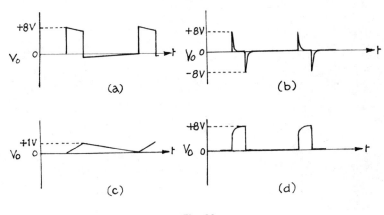

Fig. 90

ANSWERS ON DIFFERENTIATING AND INTEGRATING CIRCUITS

1 (d)	6 (a)	11 (d)	16 (c)	21 (c)
2 (a)	7 (a)	12 (a)	17 (a)	22 (c)
3 (d)	8 (d)	13 (b)	18 (b)	23 (b)
4 (d)	9 (b)	14 (b)	19 (a)	24 (c)
5 (c)	10 (c)	15 (d)	20 (d)	25 (b)

CORE STUDIES

L.C.R. CIRCUITS

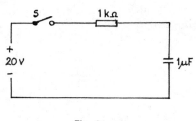

Fig. 91

(1) Refer to Fig. 91. When S is closed, the time taken for the capacitor to charge to 0.63 of the applied voltage will be:
 (a) 100ns
 (b) 1µs
 (c) 1ms
 (d) 10ms.
(2) Refer to Fig. 91. The maximum current that will flow in the circuit when S is closed will be:
 (a) 20µA
 (b) 20mA
 (c) 2µA
 (d) 20A.

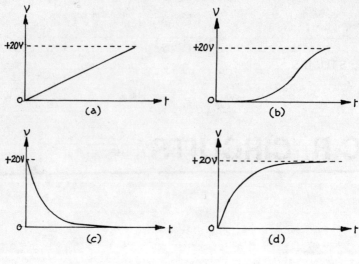

Fig. 92

(3) Refer to Fig. 91. Which of the diagrams given in Fig. 92 shows the change of voltage across the capacitor after S is closed?

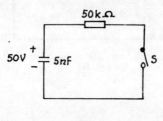

Fig. 93

(4) Refer to Fig. 93. The time taken for the voltage in the capacitor to fall to 18.5V from the instant that S is closed will be:
   (a) 250μs
   (b) 25s
   (c) 2.5ns
   (d) 2.5ms.
(5) Refer to Fig. 93. The time taken for the voltage in the capacitor to fall to 0V from the instant that S is closed will be approximately:
   (a) 0.5ms
   (b) 1.25ms
   (c) 5ms
   (d) 5s.

L.C.R. CIRCUITS 81

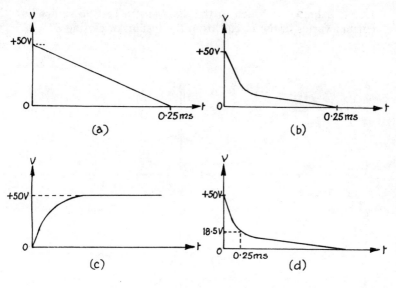

Fig. 94

(6) Refer to Fig. 93. Which of the diagrams given in Fig. 94 shows how the voltage across the capacitor varies from the instant that S is closed?

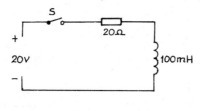

Fig. 95

(7) Refer to Fig. 95. The time taken for the current in the circuit to reach 0.63A after S is closed will be:
(a) 5ms
(b) 2s
(c) 2ms
(d) 500μs.

(8) Refer to Fig. 95. The maximum current that will flow in the circuit will be:
(a) 0.63A
(b) 0.37A
(c) 1A
(d) 5mA.

(9) Refer to Fig. 95. Which of the diagrams in Fig. 96 shows how the current varies in the circuit from the instant of closing S?

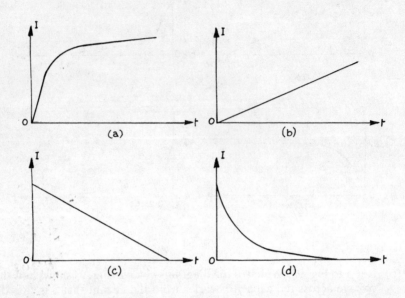

Fig. 96

(10) Refer to Fig. 97. Which of the diagrams shows an example of a **high pass filter**?

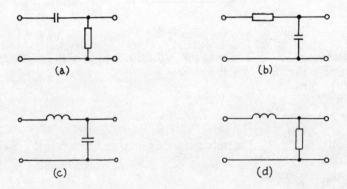

Fig. 97

(11) Refer to Fig. 98. Which of the diagrams shows an example of a low pass filter?

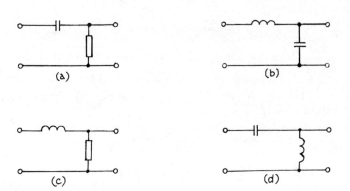

Fig. 98

(12) The **reactance** of a capacitor is given by:
  (a) $X = \pi c$
  (b) $X = 2\pi fc$
  (c) $X = \frac{1}{2\pi fc}$
  (d) $X = \frac{1}{\pi fc}$.

(13) The **reactance** of an inductor is given by:
  (a) $X = 4\pi fL$
  (b) $X = 2\pi fL$
  (c) $X = \frac{1}{2\pi fL}$
  (d) $X = \pi fL$.

(14) The reactance of a 0.002µF capacitor at a frequency of 5kHz will be approximately:
  (a) 16 k-ohm
  (b) 1.6 k-ohm
  (c) 160 k-ohm
  (d) 6 M-ohm.

(15) The reactance of 10µH inductor at a frequency of 1MHz will be approximately:
  (a) 300 ohm
  (b) 3 k-ohm
  (c) 33 k-ohm
  (d) 60 ohm.

84   ELECTRONICS SERVICING – QUESTIONS FOR PART 2

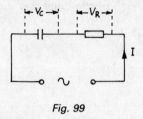

Fig. 99

(16) Refer to Fig. 99. Which of the diagrams in Fig. 100 shows the correct phasor diagram?

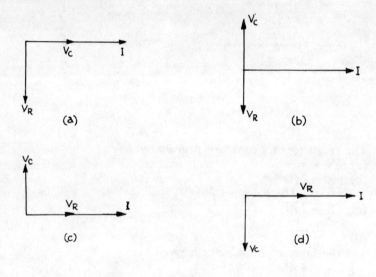

Fig. 100

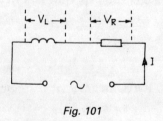

Fig. 101

(17) Refer to Fig. 101. Which of the diagrams in Fig. 102 shows the correct phasor diagram?

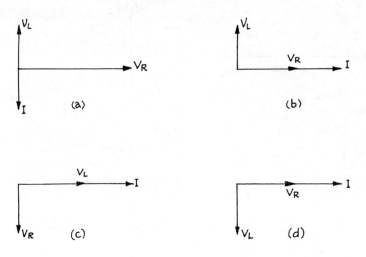

Fig. 102

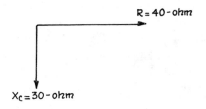

Fig. 103

(18) Refer to Fig. 103. The impedance (Z) of the circuit to which the phasors relate will be:
(a) 10 ohm
(b) 70 ohm
(c) 1200 ohm
(d) 50 ohm.

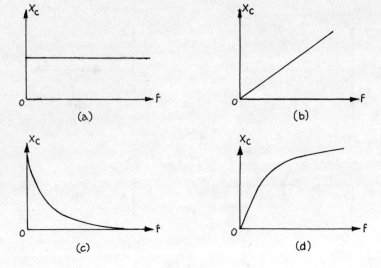

Fig. 104

(19) Which of the diagrams given in Fig. 104 shows how the reactance of a capacitor varies with frequency?

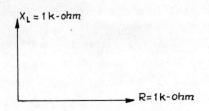

Fig. 105

(20) Refer to Fig. 105. The impedance (Z) of the circuit to which the phasors relate will be:
(a) 1.414 k-ohm
(b) 2 k-ohm
(c) Zero
(d) 0.707 k-ohm.

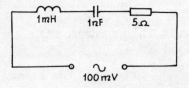

Fig. 106

L.C.R. CIRCUITS

(21) Refer to Fig. 106. The current flowing in the circuit at resonance will be:
 (a) 500mA
 (b) Zero
 (c) 20mA
 (d) 13mA.

(22) Refer to Fig. 106. The resonant frequency of the circuit will be approximately:
 (a) 159kHz
 (b) 237kHz
 (c) 230Hz
 (d) 17kHz.

(23) Refer to Fig. 106. At a frequency above resonance which of the following is true:
 (a) The inductive reactance is less than the capacitive reactance
 (b) The inductive reactance is greater than the capacitive reactance
 (c) The inductive reactance will be equal to the capacitive reactance
 (d) The circuit will be essentially resistive.

(24) Refer to Fig. 106. Which of the diagrams given in Fig. 107 shows the correct phasor diagram for circuit resonance?

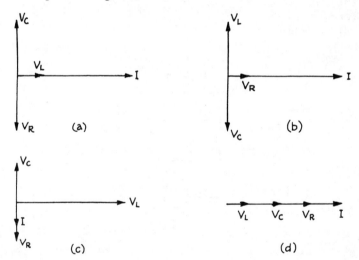

Fig. 107

(25) The resonant frequency of a series tuned circuit is given by:
 (a) $f = \pi\sqrt{LC}$
 (b) $f = 2\pi\sqrt{LC}$
 (c) $f = 4\pi\sqrt{LC}$
 (d) $f = \dfrac{1}{2\pi\sqrt{LC}}$.

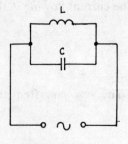

Fig. 108

(26) Refer to Fig. 108. Which of the following is true:
 (a) At resonance the impedance is purely resistive and low
 (b) At resonance the impedance is largely capacitive
 (c) At resonance the circuit is drawing its maximum current
 (d) At resonance the impedance is purely resistive and high.

(27) Refer to Fig. 108. If the values of L and C are both doubled, the resonnant frequency of the circuit will be:
 (a) Doubled
 (b) Halved
 (c) The same
 (d) Quadrupled.

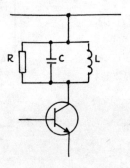

Fig. 109

(28) Refer to Fig. 109. The circuit will have its largest bandwidth when the value of R is:
 (a) 1 k-ohm
 (b) 1 M-ohm
 (c) 560 ohm
 (d) 2.2 M-ohm.

(29) Refer to Fig. 109. the selectivity of the circuit will be at its best when the value of R is:
(a) 5.6 k-ohm
(b) 1 M-ohm
(c) 560 k-ohm
(d) 2.3 k-ohm.

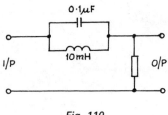

Fig. 110

(30) Refer to Fig. 110. Which of the following frequency signal voltages will result in the smallest output when applied at the input of the circuit:
(a) 2kHz
(b) 5kHz
(c) 15kHz
(d) 17kHz.

ANSWERS ON LCR CIRCUITS

1 (c)	6 (d)	11 (b)	16 (d)	21 (c)	26 (d)
2 (b)	7 (a)	12 (c)	17 (b)	22 (a)	27 (b)
3 (d)	8 (c)	13 (b)	18 (d)	23 (b)	28 (c)
4 (a)	9 (a)	14 (a)	19 (c)	24 (b)	29 (b)
5 (b)	10 (a)	15 (a)	20 (a)	25 (d)	30 (b).

CORE STUDIES

MEASURING INSTRUMENTS AND C.R.T.

(1) A moving coil instrument has a full-scale deflection of 50μA. Its sensitivity will be:
  (a) 20mV/cm
  (b) 50 k-ohm
  (c) 20 k-ohm/volt
  (d) 1 M-ohm/volt.

(2) A moving coil instrument has a full-scale deflection of 20μA. Its internal resistance on a 5V range will be:
  (a) 250 k-ohm
  (b) 100 k-ohm
  (c) 2.5 M-ohm
  (d) 1 M-ohm.

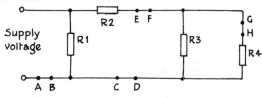

Fig. 111

(3) Refer to Fig. 111. To measure the total current drawn from the supply an ammeter could be inserted in series between points:
  (a) A and B
  (b) E and F
  (c) C and D
  (d) G and H.

(4) Refer to Fig. 111. The supply voltage is given by the voltage drops across:
  (a) R2 + R1
  (b) R1 + R2 + R3
  (c) R3 + R4 + R1
  (d) R4 + R2.

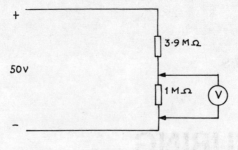

Fig. 112

(5) Refer to Fig. 112. The voltmeter has a sensitivity of 50 k-ohm/volt. On its 10V range it will read about:
(a) 3.9V
(b) 10V
(c) 5V
(d) 3.3V.

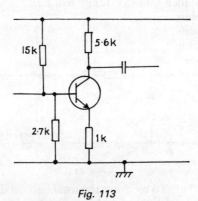

Fig. 113

(6) Refer to Fig. 113. The base-emitter junction has a forward resistance of 1.2 k-ohm and a reverse resistance of 2 M-ohm. When an ohmmeter is connected between the base and chassis reversing the polarity of its leads, it will read about:
(a) 2.7 k-ohm one way and 2 M-ohm the other way
(b) 1.2 k-ohm one way and 2.7 k-ohm the other way
(c) 2.7 k-ohm both ways
(d) 0.8 k-ohm one way and 2.5 k-ohm the other way.

MEASURING INSTRUMENTS AND C.R.T

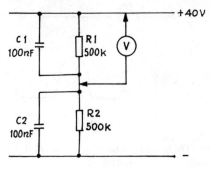

Fig. 114

(7) Refer to Fig. 114. A reading of 40V on the voltmeter probably indicates that:
(a) C2 is open-circuit
(b) R1 is open-circuit
(c) R2 is open-circuit
(d) C2 is short-circuit.

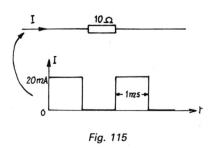

Fig. 115

(8) Refer to Fig. 115. A moving coil voltmeter connected across the resistor will indicate:
(a) 100mV
(b) 200mV
(c) 20mV
(d) 2mV.

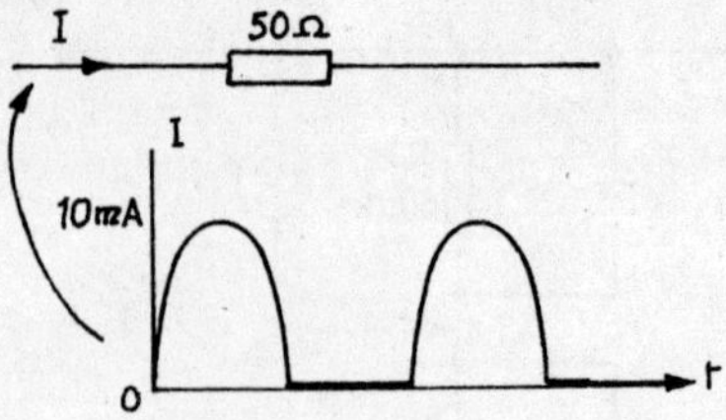

Fig. 116

(9) Refer to Fig. 116. A moving coil voltmeter connected across the resistor will indicate approximately:
   (a) 320mV
   (b) 160mV
   (c) 80mV
   (d) 500mV.

(10) The input impedance of an electronic voltmeter will most probably be:
   (a) 5 M-ohm
   (b) 200 k-ohm
   (c) 20 k-ohm
   (d) 2 k-ohm.

(11) A digital voltmeter will contain:
   (a) A moving coil movement
   (b) A moving iron movement
   (c) An A-D converter
   (d) A D-A converter.

MEASURING INSTRUMENTS AND C.R.T

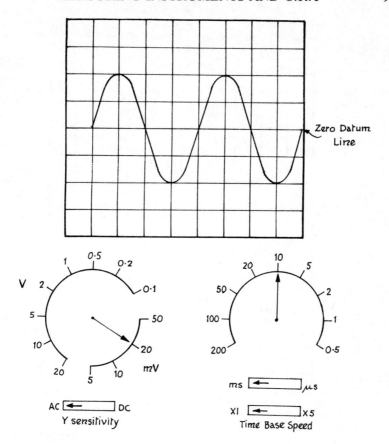

Fig. 117

(12) Refer to Fig. 117. The peak value of the waveform displayed is:
   (a) 80mV
   (b) 40mV
   (c) 20mV
   (d) 20V.

(13) Refer to Fig. 117. The frequency of the waveform displayed is:
   (a) 25Hz
   (b) 100kHz
   (c) 50Hz
   (d) 12.5Hz.

(14) Refer to Fig. 117. The r.m.s. value of the waveform displayed will be about:
   (a) 56mV
   (b) 113mV
   (c) 12mV
   (d) 28mV.

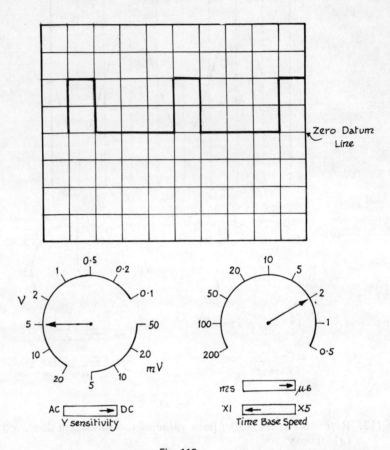

Fig. 118

(15) Refer to Fig. 118. The p.r.f. of the displayed waveform is:
 (a) 100kHz
 (b) 500kHz
 (c) 125kHz
 (d) 500Hz.
(16) Refer to Fig. 118. The d.c. component of the displayed waveform is:
 (a) +2.5V
 (b) +3.3V
 (c) 0V
 (d) +10V.

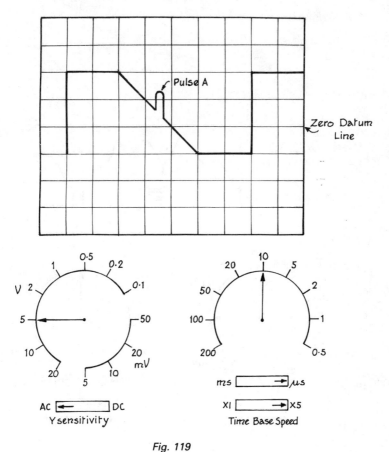

Fig. 119

(17) Refer to Fig. 119. The duration of pulse A is approximately:
 (a) 3ms
 (b) 0.6μs
 (c) 3μs
 (d) 5μs.
(18) Refer to Fig. 119. The amplitude of pulse A is about:
 (a) 2V
 (b) 5V
 (c) 15V
 (d) 10V.

98 ELECTRONICS SERVICING – QUESTIONS FOR PART 2

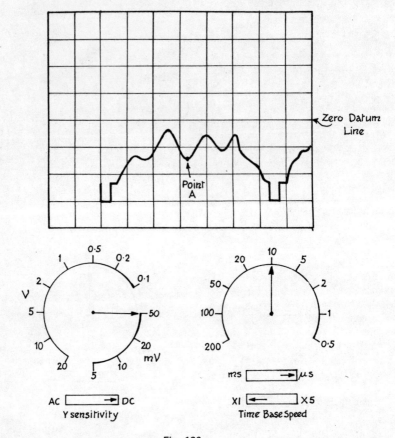

Fig. 120

(19) Refer to Fig. 120. The frequency of the displayed waveform is approximately:
 (a) 15.6kHz
 (b) 65kHz
 (c) 10kHz
 (d) 32kHz.
(20) Refer to Fig. 120. The peak-to-peak amplitude of the waveform is about:
 (a) 130V
 (b) 65V
 (c) 130mV
 (d) 65mV.

MEASURING INSTRUMENTS AND C.R.T

(21) Refer to Fig. 120. The d.c. level at point A of the waveform is about:
 (a) + 75mV
 (b) − 75mV
 (c) + 75V
 (d) − 25mV.

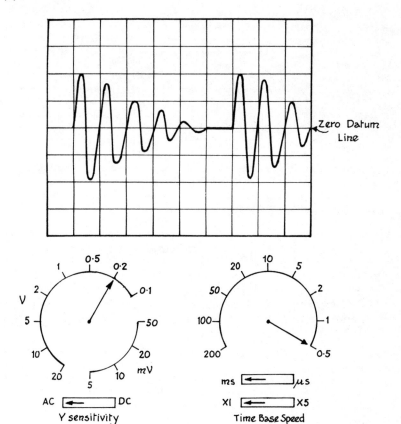

Fig. 121

(22) Refer to Fig. 121. The frequency of the damped oscillation is:
 (a) 1MHz
 (b) 2kHz
 (c) 200Hz
 (d) 500Hz.
(23) Refer to Fig. 121. The damped oscillation repeats itself every:
 (a) 6ms
 (b) 3ms
 (c) 1.2ms
 (d) 12μs.

100 ELECTRONICS SERVICING – QUESTIONS FOR PART 2

(24) Refer to Fig. 121. The maximum peak-to-peak amplitude of the waveform is about:
 (a) 0.6V
 (b) 0.4V
 (c) 750mV
 (d) 1.8V.

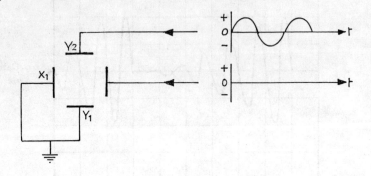

Fig. 122

(25) Refer to Fig. 122 which shows the voltage waveforms applied to the X and Y plates of a c.r.t. Which of the waveforms in Fig. 123 will be displayed on the c.r.t. screen?

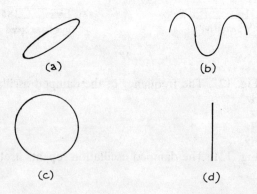

Fig. 123

MEASURING INSTRUMENTS AND C.R.T 101

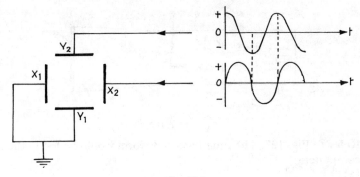

Fig. 124

(26) Refer to Fig. 124 which shows the voltage waveforms applied to the X and Y plates of a c.r.t. Which of the waveforms given in Fig. 125 will be displayed on the c.r.t. screen?

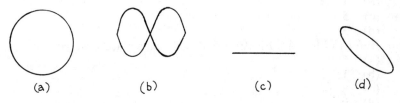

Fig. 125

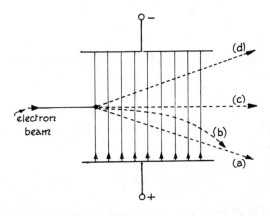

Fig. 126

(27) Refer to Fig. 126 which shows an electron beam approaching the deflector plates inside a c.r.t. Which of the dotted lines shows the path taken by the electrons?

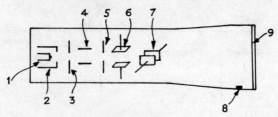

Fig. 127 C.R.T.

(28) Refer to Fig. 127. The time-base waveform would be fed to the c.r.t. component:
   (a) 3
   (b) 7
   (c) 6
   (d) 4.

(29) Refer to Fig. 127. The 1st anode is component:
   (a) 8
   (b) 1
   (c) 5
   (d) 3.

(30) Refer to Fig. 127. Electron emission takes place from component:
   (a) 6
   (b) 1
   (c) 8
   (d) 9.

(31) Refer to Fig. 127. Aluminising would be found on component:
   (a) 8
   (b) 1
   (c) 5
   (d) 9.

(32) Refer to Fig. 127. The P.D.A. electrode is component:
   (a) 5
   (b) 8
   (c) 3
   (d) 1.

(33) Refer to Fig. 127. To intensity modulate the electron beam a signal would be fed to component:
   (a) 6
   (b) 7
   (c) 3
   (d) 2.

(34) Refer to Fig. 127. The focus variable voltage would be fed to component:
   (a) 8
   (b) 2
   (c) 4
   (d) 9.

(35) Refer to Fig. 127. The waveform to be displayed would be fed to component:
- (a) 6
- (b) 8
- (c) 7
- (d) 4.

(36) In a television tube deflection of the electron beam is provided by a:
- (a) Magnetic field
- (b) Electric field
- (c) Electromagnetic field
- (d) Electrostatic field.

(37) The final anode voltage in a television tube would be about:
- (a) 500V
- (b) 50V
- (c) 5kV
- (d) 15kV.

(38) In a television tube, the video signal is normally applied to the:
- (a) Final anode
- (b) Accelerating anode
- (c) Cathode
- (d) Scan coils.

(39) The internal and external coatings of a television tube form a capacitor of value about:
- (a) 1.0μF
- (b) 0.1μF
- (c) 1pF
- (d) 750pF.

ANSWERS ON MEASURING INSTRUMENTS AND C.R.T.

1 (c)	11 (c)	21 (b)	31 (d)
2 (a)	12 (b)	22 (b)	32 (b)
3 (a)	13 (a)	23 (b)	33 (d)
4 (d)	14 (d)	24 (c)	34 (c)
5 (a)	15 (c)	25 (d)	35 (a)
6 (b)	16 (a)	26 (a)	36 (a)
7 (d)	17 (b)	27 (b)	37 (d)
8 (a)	18 (b)	28 (b)	38 (c)
9 (b)	19 (a)	29 (d)	39 (d)
10 (a)	20 (c)	30 (b)	

CORE STUDIES

TRANSFORMERS AND SHIELDING

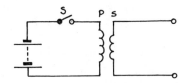

Fig. 128

(1) Refer to Fig. 128. A voltage will appear across the secondary winding:
    (a)   During the time that S is held closed
    (b)   Briefly and only when S closes
    (c)   Briefly and only when S opens or closes
    (d)   During the time that S is held open.

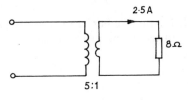

Fig. 129

(2) Refer to Fig. 129. The current flowing in the primary will be:
    (a)   500mA
    (b)   12.5A
    (c)   62.5A
    (d)   100mA.

(3) Refer to Fig. 129. The voltage across the primary winding will be:
    (a)   20V
    (b)   4V
    (c)   200V
    (d)   100V.

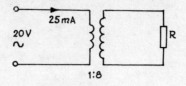

Fig. 130

(4) Refer to Fig. 130. The voltage induced into the secondary winding will be:
(a) 160V
(b) 20V
(c) 2.5V
(d) 1280V.

(5) Refer to Fig. 130. The current flowing in R will be:
(a) 3.125mA
(b) 200mA
(c) 2A
(d) 1.6A.

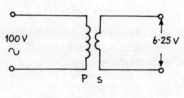

Fig. 131

(6) Refer to Fig. 131. The number of turns on the secondary is 64 turns. The number of turns on the primary will be:
(a) 4
(b) 1024
(c) 6400
(d) 16384.

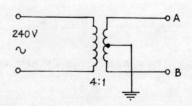

Fig. 132

(7) Refer to Fig. 132. The voltage between A and B will be:
 (a) 120V
 (b) 60V
 (c) 960V
 (d) 15V.
(8) Refer to Fig. 132. The voltage between A and earth will be:
 (a) 60V
 (b) 15V
 (c) 480V
 (d) 30V.

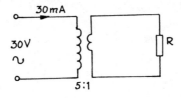

Fig. 133

(9) Refer to Fig. 133. The power dissipated in R will be:
 (a) 0.18W
 (b) 0.9W
 (c) 4.5W
 (d) 1.0W.
(10) Refer to Fig. 133. The current flowing in R will be:
 (a) 1.2mA
 (b) 6mA
 (c) 150mA
 (d) 750mA.

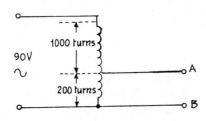

Fig. 134

(11) Refer to Fig. 134. The voltage between A and B will be:
 (a) 18V
 (b) 450V
 (c) 3.6V
 (d) 15V.

108 ELECTRONICS SERVICING – QUESTIONS FOR PART 2

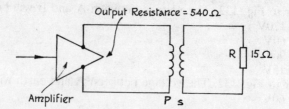

Fig. 135

(12) Refer to Fig. 135. Maximum power will be dissipated in R when the turns ratio (primary:secondary) of the transformer is:
(a) 36:1
(b) 6:1
(c) 1:36
(d) 9:1.

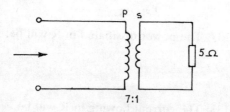

Fig. 136

(13) Refer to Fig. 136. The resistance seen 'looking' into the primary winding will be:
(a) 245 ohm
(b) 0.7 ohm
(c) 35 ohm
(d) 5 ohm.

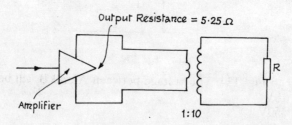

Fig. 137

(14) Refer to Fig. 137. To transfer maximum power from the amplifier to R, the value of R should be:
 (a) 0.0525 ohm
 (b) 52.5 ohm
 (c) 525 ohm
 (d) 5,250 ohm.

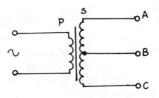

Fig. 138

(15) Refer to Fig. 138. The secondary is a continuous winding. With respect to B the voltages at A and C are:
 (a) 90° out of phase
 (b) 180° out of phase
 (c) In-phase
 (d) 270° out of phase.
(16) Refer to Fig. 138. The secondary is a continuous winding. With respect to C the voltages at A and B are:
 (a) In-phase
 (b) 90° out of phase
 (c) 180° out of phase
 (d) 270° out of phase.

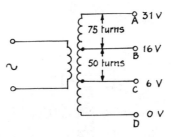

Fig. 139

(17) Refer to Fig. 139. The number of turns between C and D will be:
 (a) 20
 (b) 25
 (c) 30
 (d) 80.

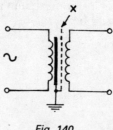

Fig. 140

(18) Refer to Fig. 140. The purpose of component X is:
(a) To increase the magnetic coupling of the two windings
(b) To decrease the magnetic coupling of the two windings
(c) To reduce the reluctance of the magnetic circuit
(d) To reduce the capacitive coupling of the two windings.

(19) Refer to Fig. 140. The iron core is used to:
(a) Reduce vibration
(b) Reduce the inductance of the windings
(c) Ensure maximum linking of flux between windings
(d) Quickly conduct heat away from the windings.

(20) The iron core of a transformer is usually laminated to:
(a) Decrease the flow of eddy currents
(b) Reduce weight
(c) Increase the hysterisis loss
(d) Reduce the number of turns used.

(21) A suitable material for use as a magnetic screen for an r.f. transformer would be:
(a) Iron
(b) Nichrome
(c) Steel
(d) Copper.

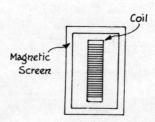

Fig. 141 Low frequency magnetic screen.

(22) Refer to Fig. 141. Which of the diagrams given in Fig. 142 shows the correct flux path?

TRANSFORMERS AND SHIELDING

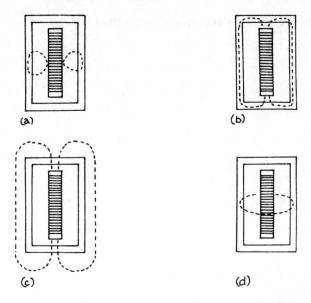

Fig. 142

(23) Refer to Fig. 143. Which of the positions for the core will result in the largest inductance of the coil?

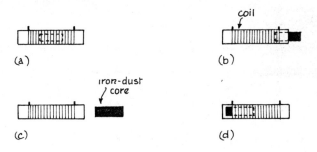

Fig. 143

ANSWERS ON TRANSFORMERS AND SHIELDING

1 (c)	6 (b)	11 (d)	16 (a)	21 (d)
2 (a)	7 (b)	12 (b)	17 (c)	22 (b)
3 (d)	8 (d)	13 (a)	18 (d)	23 (a)
4 (a)	9 (b)	14 (c)	19 (c)	
5 (a)	10 (c)	15 (b)	20 (a)	

TELEVISION AND RADIO RECEPTION

(1) Primary light sources are placed at the corners of a triangle and projected inwards as in Fig. 144.

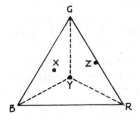

Fig. 144 Colour triangle.

What will be the colours observed at points X, Y and Z?
(2) The standard white used in colour television is:
    (a)   Illuminant A
    (b)   Illuminant B
    (c)   Illuminant C
    (d)   Illuminant D.
(3) Cyan is the complementary colour of:
    (a)   Green
    (b)   Red
    (c)   Yellow
    (d)   Blue.
(4) When cyan and magenta lights are mixed the result will be:
    (a)   Desaturated blue
    (b)   White
    (c)   Black
    (d)   Pink.

(5) When white light is added to a pure hue, the result may be described as:
   (a) Fully saturated
   (b) Desaturated
   (c) Complementary
   (d) Weighted.

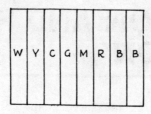

Fig. 145 Colour bars 100% AMP. 100% SAT.

(6) Refer to Fig. 145. If the blue gun is switched-off, what will be the colours displayed on the screen?

(7) Refer to Fig. 145. Determine the luminance value for each bar with the blue gun switched-off, using the formula
$$Ey = 0.59Eg + 0.3Er + 0.11Eb.$$

(8) Sketch the primary signal waveform applied to the green gun cathode during the display of the colour bars shown in Fig. 145.

(9) Interlaced scanning is used in television internationally to:
   (a) Reduce the line frequency
   (b) Reduce flicker and to avoid doubling the bandwidth
   (c) Increase the number of scanning lines used
   (d) Improve the horizontal definition.

(10) An odd number of scanning lines is used in television to:
   (a) Improve the vertical resolution
   (b) Reduce the line flyback time
   (c) Prevent picture fold-over
   (d) Permit interlaced scanning.

(11) A television system using 525 lines transmits 30 pictures per second. Its line frequency will be:
   (a) 555Hz
   (b) 15625Hz
   (c) 15750Hz
   (d) 10125Hz.

(12) The television line frequency used in this country is:
   (a) 625Hz
   (b) 15625Hz
   (c) 50Hz
   (d) 15kHz.

TELEVISION AND RADIO RECEPTION

(13) The field frequency of a t.v. broadcast transmitting 25 pictures/sec will be:
  (a) 25Hz
  (b) 50Hz
  (c) 100Hz
  (d) 12.5Hz.

(14) A 625 line television broadcast (vision and sound) occupies a channel bandwidth of:
  (a) 5.5MHz
  (b) 8MHz
  (c) 6MHz
  (d) 88MHz.

(15) The frequency spacing between vision and sound carriers in a 625 line t.v. broadcast is:
  (a) 6MHz
  (b) 8MHz
  (c) 1.57MHz
  (d) 5.5MHz.

(16) The frequency spacing between the sound and chrominance carriers in a 625 line t.v. transmission is:
  (a) 4.43MHz
  (b) 6MHz
  (c) 1.25MHz
  (d) 1.57MHz.

(17) The frequency of the chrominance sub-carrier in 625 line television is:
  (a) 470MHz
  (b) 4.43MHz
  (c) 8.86MHz
  (d) 2MHz.

(18) Refer to Fig. 146. Which of the diagrams represents a vestigial sideband t.v. transmission?

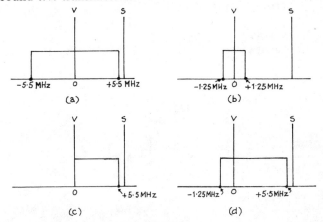

Fig. 146

(19) Bands 4 and 5 correspond to frequencies in the range of approximately:
  (a) 40 – 87MHz
  (b) 470 – 854MHz
  (c) 4 – 8 GHz
  (d) 4.4 – 8.7MHz.
(20) Refer to Fig. 147. Give the name of block A.

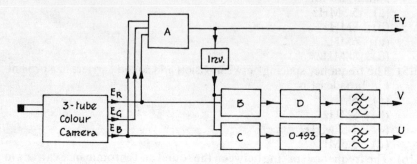

Fig. 147

(21) Refer to Fig. 147. What name is given to the signal outputs from blocks B and C?
(22) Refer to Fig. 147. What is the weighting factor applied in block D?
(23) Refer to Fig. 147. What band of frequencies is covered by:
  (1) The Ey signal?
  (2) The V and U signals?
(24) Refer to Fig. 147. Which of the output signal(s) contain the hue and saturation information of a colour picture?
(25) The ability of a colour receiver to display a good monochrome picture is a colour television system feature, referred to as:
  (a) Reverse compatibility
  (b) Duality
  (c) Compatibility
  (d) Colour tracking.

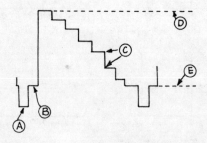

Fig. 148

(26) Refer to Fig. 148. Name the component parts A-E of the signal waveform.

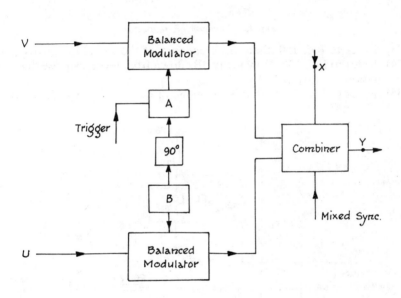

Fig. 149 Quadrature modulation principle.

(27) Refer to Fig. 149. Give the names of blocks A and B.
(28) Refer to Fig. 149. State the function of block A.
(29) Refer to Fig. 149. Name the signal at X.
(30) Refer to Fig. 149. To where is the signal at Y fed?
(31) The purpose of the colour burst is to:
    (a) Provide a high saturation level
    (b) Synchronise the receiver sub-carrier oscillator
    (c) Invert the V component
    (d) Prove hue and saturation information for the receiver.
(32) A 'swinging' burst is provided in PAL to:
    (a) Overcome phase errors in transmission
    (b) Assist in receiver a.g.c.
    (c) Correct for small errors in the receiver sub-carrier oscillator
    (d) Keep the receiver PAL switch in step.

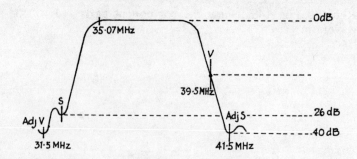

Fig. 150 Receiver I.F. response.

(33) Refer to Fig. 150. State the frequency of the sound i.f. carrier (S).
(34) Refer to Fig. 150. How many dBs down from level response does the vision i.f. carrier (V) sit?
(35) Refer to Fig. 150. What is the name of the i.f. carrier at 35.07MHz?

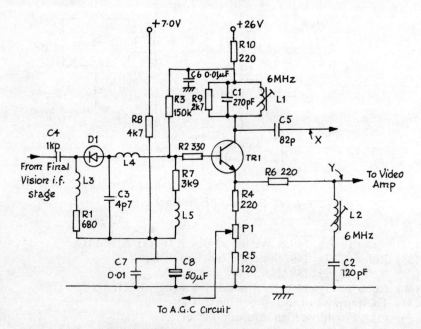

Fig. 151 Part of monochrome receiver circuit.

(36) Refer to Fig. 151. State **two** functions of D1.
(37) Refer to Fig. 151. Make a sketch of the signals present at X and Y during the reception of colour bars.
(38) Refer to Fig. 151. State the function of:
    (1) TR1
    (2) P1.

TELEVISION AND RADIO RECEPTION

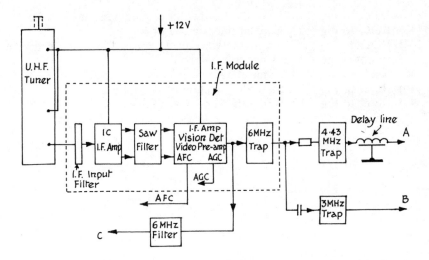

Fig. 152 Part of colour receiver circuit schematic.

(39) Refer to Fig. 152. When the receiver is tuned to a colour transmission, name the signals present at A, B and C.
(40) Refer to Fig. 152. State the frequency range present at points A, B and C during the reception of a colour transmission.
(41) Refer to Fig. 152. To where in the receiver will the a.f.c. line be fed?

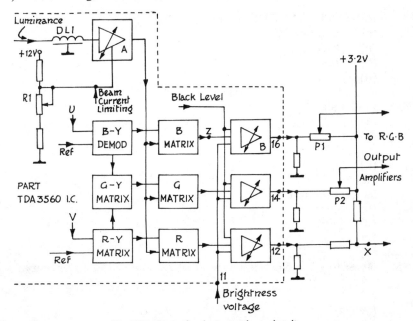

Fig. 153 Part of colour receiver circuit.

(42) Refer to Fig. 153. State the bandwidth requirement of block B.
(43) Refer to Fig. 153. Explain the purpose of R1.
(44) Refer to Fig. 153. Draw the waveform expected at point X during reception of a colour bar transmission.
(45) Refer to Fig. 153. Name the signal present at point Z.
(46) Refer to Fig. 153. State the function of P1 and P2.
(47) Refer to Fig. 153. Name the component DL1.

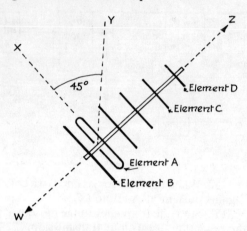

Fig. 154 Plan view of horizontal YAGI array.

(48) Refer to Fig. 154. Maximum signal pick-up will lie in:
    (a) Direction W
    (b) Direction X
    (c) Direction Y
    (d) Direction Z.
(49) Refer to Fig. 154. The coaxial cable down lead is normally connected to element:
    (a) A
    (b) B
    (c) C
    (d) D.
(50) Refer to Fig. 154. For operation at 600MHz, element A would have a length of about:
    (a) 10cm
    (b) 0.06m
    (c) 25cm
    (d) 50cm.
(51) Refer to Fig. 154. Minimum signal pick-up will lie in direction:
    (a) W
    (b) X
    (c) Y
    (d) Z.

(52) Refer to Fig. 154. What plane of polarisation will be used with the aerial mounted as indicated and to which 'field' does polarisation refer?
(53) To provide a steady colour picture with little noise grain, a receiving aerial should deliver a signal to its receiver of about:
   (a) 2–5mV
   (b) 5–10μV
   (c) 100–200μV
   (d) 5–10V.

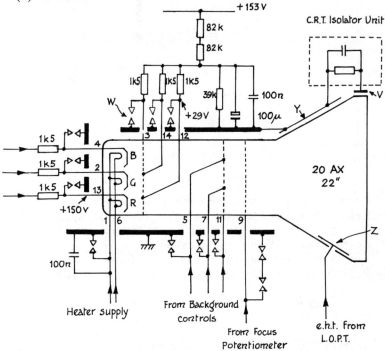

Fig. 155 In-line colour C.R.T.

(54) Refer to Fig. 155. Name the components V, Y and Z.
(55) Refer to Fig. 155. Determine the bias voltage on the red gun.
(56) Refer to Fig. 155. State the function of component W.
(57) Refer to Fig. 155. Name the stage of the receiver to which pin 2 is connected.
(58) Refer to Fig. 155. The voltage fed to pin 9 would be about:
   (a) +200V
   (b) −200V
   (c) +600V
   (d) +4.5kV.
(59) Refer to Fig. 155. State the function of the C.R.T. isolator.
(60) Refer to Fig. 155. State the purpose of the controls connected to pins 5, 7 and 11.

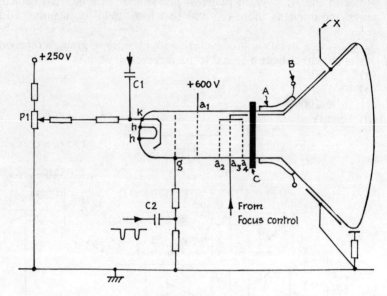

Fig. 156 Monochrome C.R.T.

(61) Refer to Fig. 156. Name the components A, B and C.
(62) Refer to Fig. 156. State the purpose of the pulses applied to C2.
(63) Refer to Fig. 156. State the effect on the picture if the voltage present on the slider of P1 increases from +80V to +90V.
(64) Refer to Fig. 156. Sketch the signal waveform applied to C1.
(65) Refer to Fig. 156. The voltage applied to X would be about:
   (a) + 50V
   (b) + 18kV
   (c) + 4kV
   (d) 0V.
(66) Refer to Fig. 157. Name the type of synchronisation used for the line oscillator.
(67) Refer to Fig. 157. Sketch the waveforms present at W and X.
(68) Refer to Fig. 157. State the purpose of block C.
(69) Refer to Fig. 157. Sketch the waveform at Y during the field sync. interval.
(70) Refer to Fig. 157. What will be the probable effect on the picture caused by adjustment of R1?
(71) Refer to Fig. 158. State the purpose of the voltage present at X and Y.
(72) Refer to Fig. 158. Sketch the waveform present at Z.
(73) Refer to Fig. 158. State the function of D1, C1.
(74) Refer to Fig. 158. Give the purpose of winding w_1.
(75) Refer to Fig. 158. State the effect on the picture if the +125V suply is lower than normal.
(76) Refer to Fig. 158. Name the function performed by C2.

TELEVISION AND RADIO RECEPTION

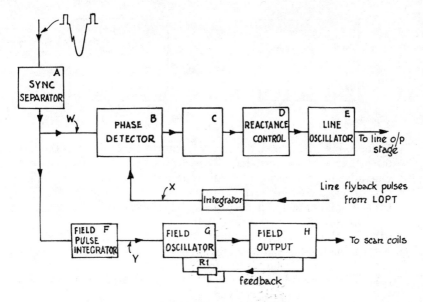

Fig. 157 Timebase synchronisation.

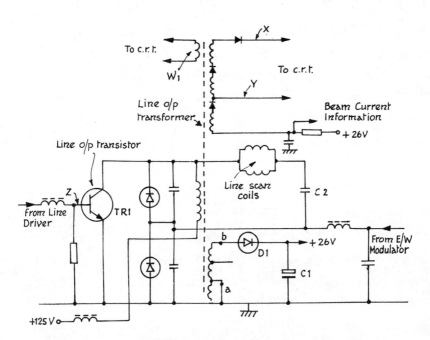

Fig. 158 Part line output stage.

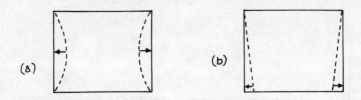

Fig. 159

(77) Refer to Fig. 159. Name the receiver control used to correct the raster distortions (shown dotted).
(78) One advantage of using a tuned r.f. amplifier in a v.h.f./f.m. radio is:
 (a) The stage bandwidth is broadened
 (b) A smaller signal is delivered to the mixer
 (c) Less noise is introduced by the mixer
 (d) The image signal rejection is improved.
(79) An r.f. amplifier is not normally used in an a.m. radio receiver covering the L.W. and M.W. bands because:
 (a) The noise picked-up by the aerial is larger than the noise generated by the receiver
 (b) The frequency separation between transmitting stations is small
 (c) A low intermediate frequency is normally employed
 (d) It would be difficult to match to the ferrite rod aerial.

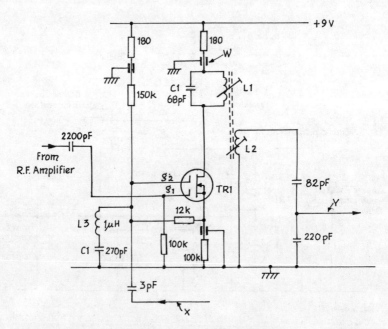

Fig. 160 Part of F.M. radio receiver.

(80) Refer to Fig. 160. To which stage in the receiver would point X be connected?
(81) Refer to Fig. 160. Name the component W.
(82) Refer to Fig. 160. State the frequency to which L1, C1 would be tuned and the 3dB bandwidth of the tuned circuit.
(83) Refer to Fig. 160. To which stage in the receiver would point Y be connected?
(84) Refer to Fig. 160. What is the function of TR1?

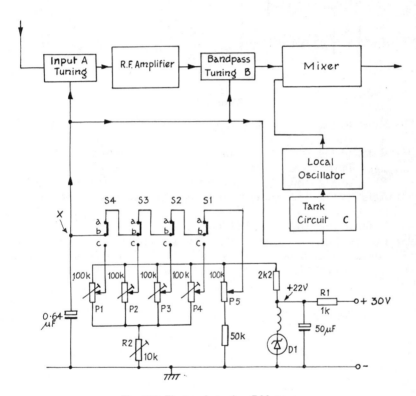

Fig. 161 Electronic tuning F.M. tuner.

(85) Refer to Fig. 161. State the purpose of D1.
(86) Refer to Fig. 161. State the type of voltage to be found at point X.
(87) Refer to Fig. 161. State how the tuning voltage alters the frequency of operation in blocks A, B and C.
(88) Refer to Fig. 161. State which component is used for tuning to any station within the entire f.m. band.
(89) Refer to Fig. 161. How is the tuning point controlled when S2 is operated?

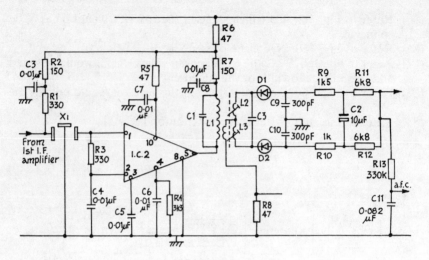

Fig. 162 Part F.M. receiver.

(90) Refer to Fig. 162. Which component(s) determine the selectivity of the circuit shown?

(91) Refer to Fig. 162. Which component(s) provide the gain to the i.f. signal in the circuit shown?

(92) Refer to Fig. 162. Which component(s) are used for matching X1?

(93) Refer to Fig. 162. State the function of the circuitry associated with D1, D2.

(94) Refer to Fig. 162. State the function of C2.

(95) Refer to Fig. 162. A signal of 10.7MHz at a suitable level is injected at pin 1 of I.C.2. State how the voltage across C11 will be affected as the injected signal is varied between 10.6MHz and 10.8MHz.

ANSWERS ON TELEVISION AND RADIO RECEPTION

1. X- desaturated cyan, Y- white and Z-yellow.
2. (d)
3. (b)
4. (a)
5. (b)
6. From left to right   yellow, yellow, green, green, red, red, black, black.
7. From left to right   0.89, 0.89, 0.59, 0.59, 0.3, 0.3, 0.0, 0.0.
8.

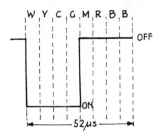

Fig. 163

9. (b)
10. (d)
11. (c)
12. (b)
13. (b)
14. (b)
15. (a)
16. (d)
17. (b)
18. (d)
19. (b)
20. Luminance signal matrix.
21. Colour difference signals (Er-Ey from B and Eb-Ey from C).
22. 0.877.
23. (1) 0–5.5MHz, (2) 0–1MHz.
24. The V and U signals.
25. (a)
26. A – line sync. pulse, B – back porch, C – luminance signal (grey scale), D – peak white level, E – black and blanking level.
27. A – PAL switch, B – sub-carrier (reference) oscillator (4.43MHz).
28. To provide inversion of the V component on alternate lines.
29. Luminance signal.
30. To the main u.h.f. carrier modulator.
31. (b)
32. (d)

128   ELECTRONICS SERVICING – QUESTIONS FOR PART 2

33   33.5MHz (6MHz below the vision i.f.).
34   6dB.
35   Chrominance i.f. carrier.
36   (a)   To demodulate the video signal.
     (b)   To act as a mixer to extract the 6MHz beat (intercarrier sound) between the vision and sound i.f. carriers.
37

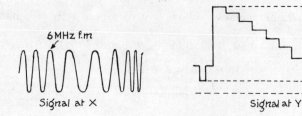

Fig. 164

38   (1) TR1 acts as a buffer stage, connected as an emitter follower to the luminance signal. Its function is to prevent the low input impedance of the video amplifier stage from loading the vision detector circuit. It also provides some gain to the 6MHz intercarrier sound signal developed across the tuned load L1, C1, operating as a common emitter amplifier to the 6MHz signal.
38   (2) The function of P1 (contrast control) is to manually vary the gain of the vision i.f. stages via the a.g.c. circuit to provide adjustment over the amplitude of the luminance signal.
39   A – luminance signal, B – chrominance signal and C – intercarrier sound signal.
40   A – 0–5.5MHz (with dip in response at 4.43MHz), B – 4.43MHz±1MHz and C – 6MHz±100kHz.
41   The a.f.c. error voltage is fed to the u.h.f. tuner to correct drift in the local oscillator. It provides the correction via the tuning voltage fed to the local oscillator vari-cap diode.
42   0–5.5MHz.
43   The purpose of R1 is to provide manual variation of the gain of the controlled amplifier A to allow contrast adjustment of the displayed picture.
44

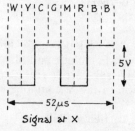

Fig. 165

TELEVISION AND RADIO RECEPTION

45 Blue primary signal.
46 P1 and P2 adjust the amplitude of the blue and green primary signals fed to their respective output stages to allow correct setting of the standard white during grey-scale adjustments (note that the red primary signal channel gain is fixed).
47 Luminance delay line (800ns).
48 (d)
49 (a)
50 (c)
51 (a)
52 Horizontal polarisation. The Electric Field.
53 (a)
54 V – rim band; Y – external aquadag coating; Z – internal coating forming final anode and connected to e.h.t. supply.
55 121V (150–29) with the grid pin 12 negative w.r.t. the cathode pin 13.
56 Component W is a spark gap and its function is to by-pass flash-over currents from the external circuits (to prevent damage) in the event of an e.h.t. flash-over within the c.r.t.
57 The green primary output video voltage amplifier.
58 (d).
59 The c.r.t. isolator removes static charges built-up on the c.r.t. rim band at a rate determined by the time-constant of the CR network. The rim band cannot be directly connected to chassis, since the chassis may be at half-mains potential or at full mains potential and accidental contact with the rim band may occur.
60 The background controls connected to these pins adjust the first anode potential of each electron gun to permit setting of the lowlights during grey-scale adjustments.
61 A – line and field scan coils; B – raster correction magnet (pin-cushion distortion); C – picture shift magnets.
62 The negative-going pulses at line and field rate are used for blanking the electron beam during line and field flyback.
63 The brightness of the picture will decrease.
64

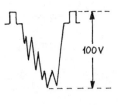

Fig. 166

65 (b)
66 Line flywheel synchronisation.

130   ELECTRONICS SERVICING – QUESTIONS FOR PART 2

67

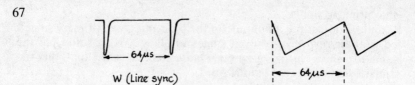

W (Line sync)

X (Integrated line flyback pulses)

Fig. 167

68   Block C is a low pass filter with values chosen so that only d.c. and low frequencies are passed to the control stage D. The low pass filter gives the arrangement its flywheel effect by smoothing out interference in the sync. waveform. The values used in the filter determines the 'pull-in' range of the circuit.

69

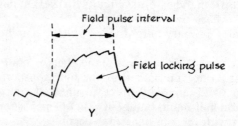

Fig. 168

70   Alteration of the vertical linearity.
71   Voltage at X provides the final anode voltage (e.h.t.) for the c.r.t. so that the electrons strike the screen at high velocity. Voltage at Y provides the focusing voltage for usually the second anode of the c.r.t. to provide a small diameter beam on striking the screen.

72

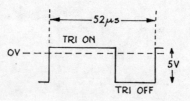

Fig. 169

73   D1 rectifies the voltage across a and b of the line output transformer to provide a low voltage supply of 26V with C1 acting as reservoir capacitor.

TELEVISION AND RADIO RECEPTION

74 The purpose of winding w_1 is to provide a voltage supply for the heaters of the c.r.t.
75 If the +125V supply is lowered, less energy will be fed into the line o/p transformer when TR1 conducts. Hence there will be less energy fed to the line scan coils and the picture width will be reduced.
76 'S' correction.
77 (a) E–W pin cushion amplitude; (b) E–W keystone amplitude.
78 (d)
79 (a)
80 Local oscillator.
81 Feed-through capacitor (low inductance).
82 10.7MHz; 200–250kHz.
83 1st i.f. amplifier.
84 TR1 functions as a multiplicative mixer stage, mixing the incoming signal frequency with the output of the local oscillator to produce a difference frequency selected by L1, C1.
85 The purpose of D1 (zener diode) in conjunction with R1 is to provide a 22V stabilised supply to the tuning potentiometers from the 30V line.
86 The voltage at point X is a d.c. voltage (the tuning voltage) varied by the tuning potentiometers.
87 The tuning voltage is applied to vari-cap diodes in blocks A, B and C which form part of the tuning capacitance. By varying the tuning voltage and hence the reverse bias applied to the vari-cap diodes the effective capacitance of the tuning circuits may be varied.
88 P5 provides manual tuning over the entire f.m. band.
89 When S2 is operated, P5 is removed from circuit and the tuning voltage (preset) is supplied from P3.
90 X1 – ceramic filter; L1, C1 and L2, L3, C3.
91 I.C.2.
92 R1 and R3.
93 F.M. detector (ratio detector – balanced).
94 Amplitude limiting of the f.m. i.f. carrier.
95 The d.c. voltage across C11 will change from a voltage of negative polarity – through zero – to a voltage of positive polarity. The voltage (a.f.c. voltage) is used to correct drift in the receiver local oscillator to maintain the correct i.f. of 10.7MHz.

INDUSTRIAL EQUIPMENT

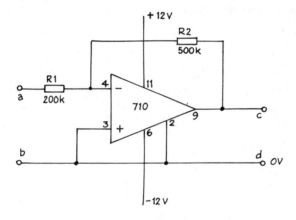

Fig. 170

(1) Refer to Fig. 170. Determine the approximate voltage gain of the amplifier.
(2) Refer to Fig. 170. A d.c. voltage of 0.5V is applied between terminals a and b with terminal a positive w.r.t. b. Determine the approximate output voltage between terminals c and d and state the polarity.
(3) Refer to Fig. 170. Is the amplifier operating under 'open loop' or 'closed loop' condition?
(4) Refer to Fig. 170. Under the circuit condition illustrated which pin of the i.c. could be referred to as a 'virtual earth'?
(5) If terminals a and b are shorted together, what voltage would be expected between terminals c and d?

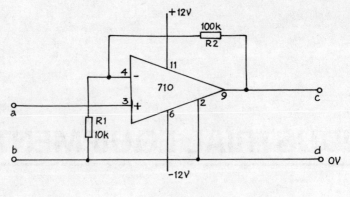

Fig. 171

(6) Refer to Fig. 171. Determine the output voltage between c and d when a 50mV r.m.s. sine wave is applied between a and b.
(7) Refer to Fig. 171. State the phase relationship between input and output signals.

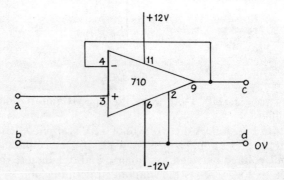

Fig. 172

(8) Refer to Fig. 172. What will be the approximate output signal voltage when a 1.4V r.m.s. sine wave is applied at the input?

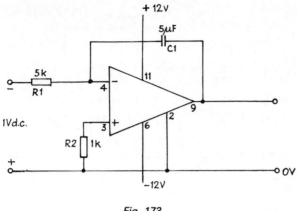

Fig. 173

(9) Refer to Fig. 173. Calculate the integrator time constant.
(10) Refer to Fig. 173. Sketch the output waveform from the circuit showing amplitude and time-scale when 1V d.c. with polarity as indicated is applied at the input (assume that C is initially uncharged).

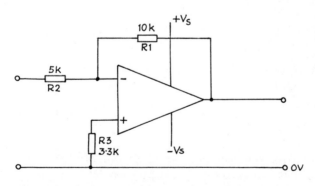

Fig. 174

(11) Refer to Fig. 174. The purpose of R3 is to:
    (a)   Reduce the amplifier gain
    (b)   Increase the CMRR
    (c)   Decrease the CMRR
    (d)   Increase the slew rate.

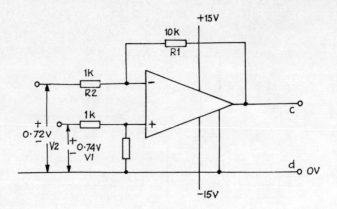

Fig. 175

(12) Refer to Fig. 175. Determine the magnitude and polarity of the output voltage from the comparator.

(13) The input offset voltage of an OP-AMP is:
   (a) The d.c. voltage applied to the input terminals to force the steady d.c. output voltage to zero
   (b) The d.c. voltage applied to the input terminals to produce an offset in the output voltage
   (c) The d.c. voltage applied to the input to force the output voltage to $+V_s$
   (d) The d.c. voltage applied to the input to force the output voltage to $-V_s$.

(14) When the inputs of an OP-AMP are held at zero, any change in the output voltage is due to:
   (a) Slewing
   (b) C.M.R.R.
   (c) Drift
   (d) Input bias current.

(15) Refer to Fig. 176. State the components that determine the duration (t) of the output pulse.

(16) Refer to Fig. 176:
   (a) Give the state of the $\overline{Q}$ output of the FF and hence the conducting state of TR1 prior to the arrival of a trigger pulse.
   (b) Give the states (high or low) at the outputs of comparators 1 and 2 prior to the arrival of a trigger pulse.
   (c) At what instant will the FF reset.

(17) Refer to Fig. 177. Given that the duration (t) of the output pulse in Fig. 176 may be found from $t = 1.1 \times C_a \cdot R_a$ secs, determine the time interval that D1 is emitting light in Fig. 177.

(18) Refer to Fig. 177. When the start button is pressed in Fig. 177, what will be the conducting state of TR1 in Fig. 176?

INDUSTRIAL EQUIPMENT

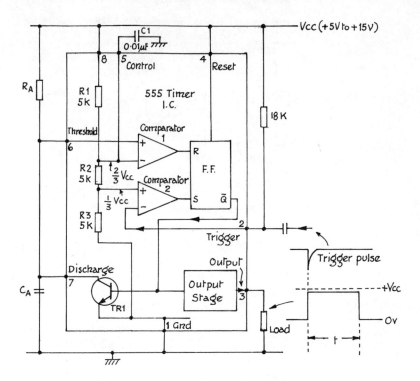

Fig. 176 555 timer I.C.

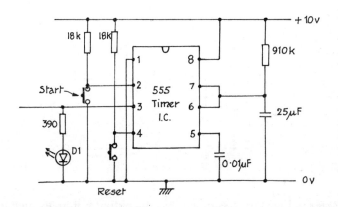

Fig. 177 Practical timer circuit.

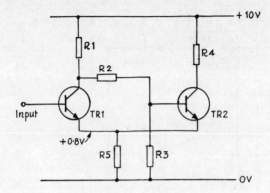

Fig. 178 Basic Schmitt trigger.

(19) Refer to Fig. 178:
   (a) If the base of TR1 is held at 0V what will be the conducting state of the two transistors.
   (b) If the base of TR1 is taken to +1.5V what will be the conducting state of the two transistors.
   (c) From where is the output of the circuit taken.

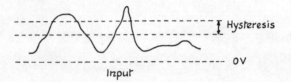

Fig. 179

(20) Refer to Fig. 178. Draw the time related output of the circuit when the waveform shown in Fig. 179 is applied at the input.

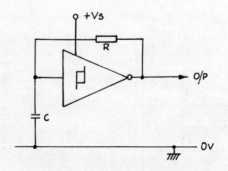

Fig. 180

(21) Refer to Fig. 180. Assuming that the output is initially high, describe the sequence of events for one cycle of oscillation.

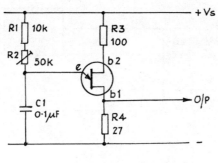

Fig. 181

(22) Refer to Fig. 181. Draw time-related waveforms for the emitter and output electrodes.
(23) A standard 74/54 TTL gate has a maximum output current in the 'low' state of 16mA. It is used to drive a number of other 74/54 TTL gates, each allowing a maximum input current in the 'low' state of -1.6mA. What is the maximum 'fan-out' of the 74/54 in the 'low' state?

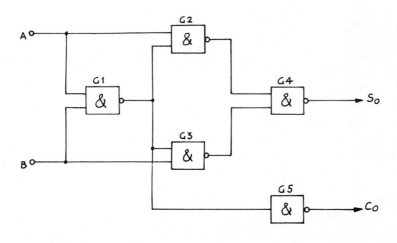

Fig. 182

(24) Refer to Fig. 182. Draw up a truth table relating the inputs A and B with the outputs So and Co.
(25) Refer to Fig. 182. State the logic function performed by the collection of gates G1–G4.

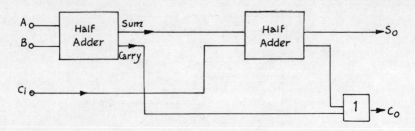

Fig. 183

(26) Refer to Fig. 183:
  (a) Name the combinational logic circuit shown.
  (b) Construct a truth table relating the inputs and outputs of the circuit.

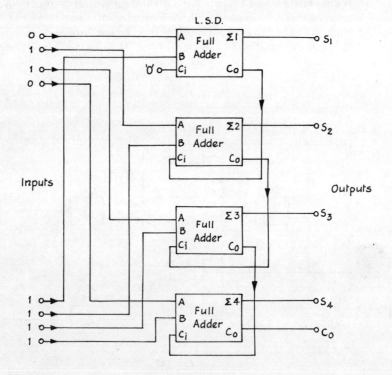

Fig. 184 Parallel adder.

(27) Refer to Fig. 184:
  (a) Give the logic state of the five outputs when the input states are as indicated.
  (b) State why the Ci input of the upper full adder is wired to logic 0.

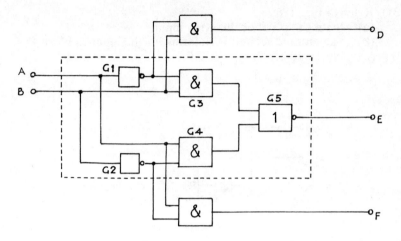

Fig. 185 1-Bit binary comparator.

(28) Refer to Fig. 185:
(a) Give the logic state at the D, E and F outputs when:
(1) A = B.
(2) A > B.
(3) A < B.
(b) State the logic function performed by the combination of gates G1–G5 and give the Boolean expression for the combination.

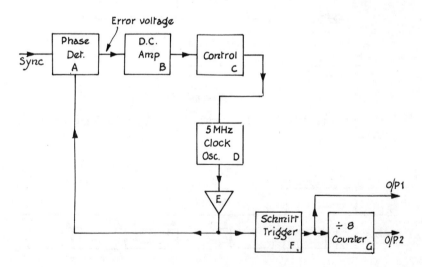

Fig. 186 Synchronised clock oscillator.

(29) Refer to Fig. 186. What type of oscillator will probably be used in D?
(30) Refer to Fig. 186. State the purpose of amplifier E.

(31) Refer to Fig. 186:
  (a) State the p.r.f. of the sync. input.
  (b) Sketch time related waveforms for the input to block F, o/p 1 and o/p 2.

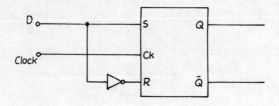

Fig. 187

(32) Refer to Fig. 187. Name the sequential logic circuit shown.
(33) Refer to Fig. 187. Draw up a truth table relating the inputs and outputs.

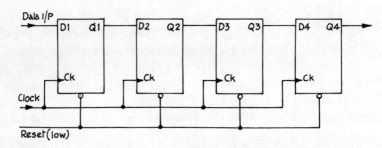

Fig. 188

(34) Refer to Fig. 188. Give the name of the sequential logic arrangement shown.

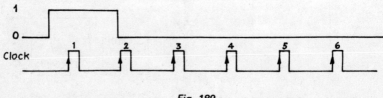

Fig. 189

(35) Refer to Fig. 188. Draw time related waveforms for the Q1, Q2, Q3 and Q4 outputs when the D1 and clock waveforms are as in Fig. 189.

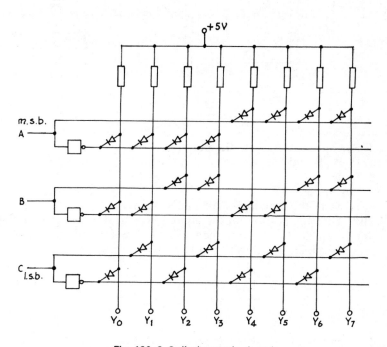

Fig. 190 3–8 diode matrix decoder.

(36) Refer to Fig. 190. Give the logic state (1 or 0) of the output lines when the binary inputs A, B and C have the following state.
(1) A = 0   (2) A = 1   (3) A = 1
     B = 1   B = 1   B = 0
     C = 1   C = 1   C = 1
Assume logic 0 = 0V and logic 1 = +5V.

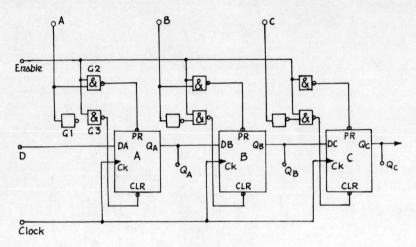

Fig. 191 Shift register.

(37) Refer to Fig. 191. State the purpose of the terminals.
    (1) A, B and C.
    (2) D.

(38) Refer to Fig. 191:
    (a) State the function of the enable terminal.
    (b) Give the purpose of the inputs PR and CLR on each flip-flop.
    (c) Explain the action when a logic 1 is applied to the enable terminal, by reference to latch A.

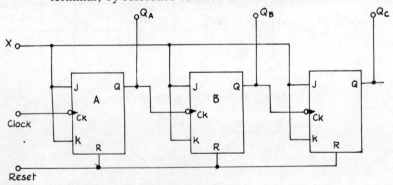

Fig. 192 Counter.

(39) Refer to Fig. 192:
    (a) Give the full name of the counter.
    (b) To where is X connected.
    (c) Give the logic state at the Qa, Qb and Qc outputs after the application of 5 clock pulses from the reset condition.
    (d) State what limits the maximum counting rate of the arrangement.

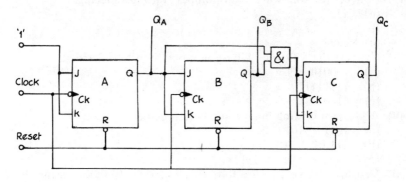

Fig. 193 Counter.

(40) Refer to Fig. 193:
- (a) Give the full name of the circuit.
- (b) Draw time-related waveforms for the clock input and the Qc output.
- (c) State what type of flip-flop is normally used in this counter.

(41) Convert the following denary numbers into hexadecimal:
- (a) 11
- (b) 256
- (c) 1026.

(42) Convert the following denary numbers into B.C.D. (natural coding):
- (a) 3459
- (b) 181
- (c) 76.

(43) Convert the following denary numbers into octal:
- (a) 10
- (b) 65
- (c) 128.

(44) Convert the following hexademical numbers into denary:
- (a) FF
- (b) 3A
- (c) 12D.

(45) Convert the following hexadecimal numbers into binary:
- (a) 118
- (b) F
- (c) 4C.

(46) Convert the following hexadecimal numbers into B.C.D. (natural coding):
- (a) 100A
- (b) 30
- (c) 50D.

146 ELECTRONICS SERVICING – QUESTIONS FOR PART 2

(47) Convert the following octal numbers into hexadecimal:
   (a) 47
   (b) 102
   (c) 2677
(48) Convert the following binary numbers into hexadecimal:
   (a) 100010
   (b) 11111111
   (c) 1100.
(49) Add the following binary numbers together:
   (a) 1011 + 1010
   (b) 111.01 + 1011.11
   (c) 1001 + 1101 + 11111.
(50 Give the answers to the following binary subtractions using the 2's complement method:
   (a) 11100101 − 10100011
   (b) 101100 − 11011
   (c) 010011 − 01110.

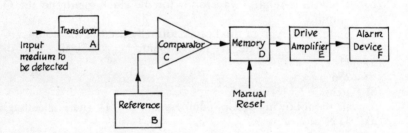

Fig. 194 Basic diagram of general alarm system.

(51) Refer to Fig. 194:
   (a) State **two** types of transducers that may be used to detect each of the following input mediums:
      (1) Temperature
      (2) Liquid level
      (3) Light level
      (4) Intruder
   (b) Give the functions of blocks B and C.
(52) Refer to Fig. 194:
   (a) State the reason for the use of block D
   (b) To gain maximum attention, what frequency would be used for an audible alarm.
(53) Refer to Fig. 195. An alarm is given when the liquid reaches the level x-y:
   (a) State the function of the gates G1 and G2 together with associated components.

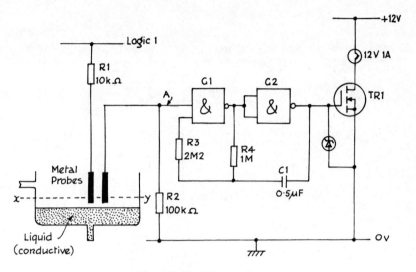

Fig. 195 Liquid level alarm.

(b) What will be the state of the lamp when:
   (1) The liquid level is as shown
   (2) The liquid reaches the level x-y.
(c) What type of device is TR1 and state any particular advantage of its use in this application.

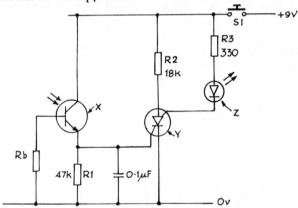

Fig. 196 Flame detector circuit.

(54) Refer to Fig. 196:
   (a) Name the devices X, Y and Z.
   (b) Briefly describe the operation of the circuit.
   (c) State the purpose of S1.
   (d) What effect on operation would a resistor Rb have when connected as shown.

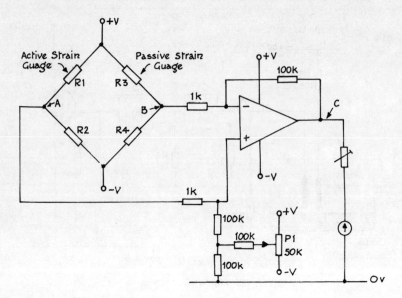

Fig. 197

(55) Refer to Fig. 197:
  (a) Explain the purpose of R3.
  (b) When $\frac{R1}{R2} = \frac{R3}{R4}$ what voltage will be present between points A and B, and at point C.
(56) Refer to Fig. 197:
  (a) State the type of amplifier used in this circuit and mention any particular advantage of its use in this application.
  (b) State the function of P1.

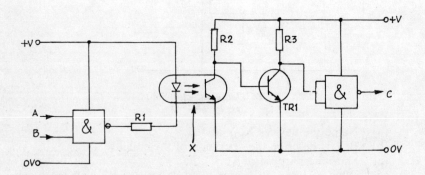

Fig. 198 Logic interface.

INDUSTRIAL EQUIPMENT

(57) Refer to Fig. 198:
  (a) Name the device X.
  (b) State any particular advantages of the device X.
(58) Refer to Fig. 198:
  (a) Give the logic state at point C ('high' or 'low') and the conducting state of TR1 for the following logic inputs to A and B:
    (1) A – 'high': B – 'high'.
    (2) A – 'high': B – 'low'.
    (3) A – 'low': B – 'low'.
  (b) State the purpose of R1.

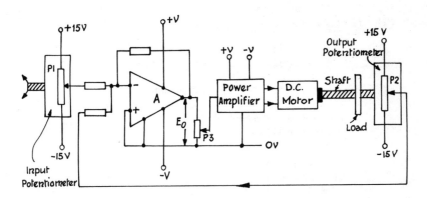

Fig. 199 Closed loop positional servo system.

(59) Refer to Fig. 199:
  (a) State the function of component A.
  (b) Assuming that P1 and P2 outputs are initially zero, explain the action when P1 slider is moved so that its output is −2V.
(60) Refer to Fig. 199:
  (a) State what is likely to happen if when Eo reaches zero, the load overshoots the desired position.
  (b) Explain the purpose of P3.

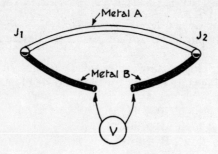

Fig. 200

(61) Refer to Fig. 200. The larger reading will be obtained on the voltmeter when:
  (a) J1 = 100°C and J2 = 0°C
  (b) J1 = 100°C and J2 = 50°C
  (c) J1 = 50°C and J2 = 100°C
  (d) J1 = 100°C and J2 = 100°C

(62) Refer to Fig. 200. The voltmeter will read zero when:
  (a) J1 = 0°C and J2 = 5°C
  (b) J1 = 50°C and J2 = 50°C
  (c) J1 = 0°C and J2 = 100°C
  (d) J1 = 5°C and J2 = 50°C.

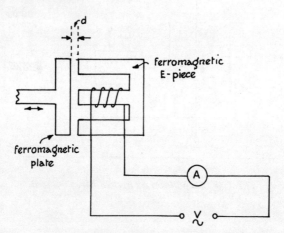

Fig. 201

(63) Refer to Fig. 201:
  (a) State the name of the transducer.
  (b) Explain how the current reading is affected when the distance d is increased.
  (c) What will be the phase relationship between voltage and current in the circuit.

INDUSTRIAL EQUIPMENT

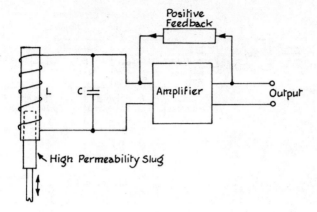

Fig. 202

(64) Refer to Fig. 202:
    (a) State the name of the transducer used in the circuit.
    (b) Explain how the output will be affected when the slug is moved further into the inductor.

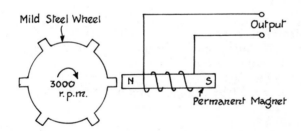

Fig. 203

(65) Refer to Fig. 203:
    (a) Name the transducer, the principle of which is illustrated by the diagram.
    (b) State the frequency of the output voltage.

(66) Refer to Fig. 203:
    (a) State the relationship between the rotary speed of the wheel and the output voltage.
    (b) State the effect if the direction of rotation is reversed.

152 ELECTRONICS SERVICING – QUESTIONS FOR PART 2

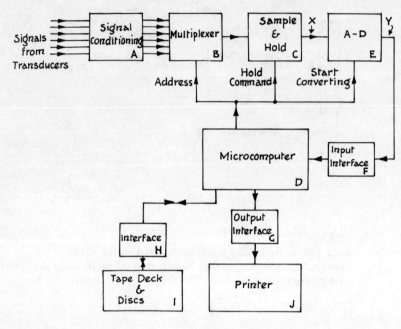

Fig. 204

(67) Refer to Fig. 204:
  (a) Name the system under the control of the microcomputer.
  (b) State the function of block B.
(68) Refer to Fig. 204:
  (a) State the type of signals present at points X and Y.
  (b) What is the function of block C.
(69) Refer to Fig. 204:
  (a) Which block(s) may hold a permanent record of transducer data.
  (b) When the systems software programme is loaded from tape/disc, name the area inside the microcomputer that will hold the programme.
  (c) Give the purpose of block A.
(70) Refer to Fig. 205. Draw time-related waveforms showing the voltage across the load when the gate trigger is at x and at y.
(71) Refer to Fig. 206:
  (a) State the function of the flywheel diode D1.
  (b) Draw the current waveshape (I) in the load with and without the diode in circuit.

INDUSTRIAL EQUIPMENT

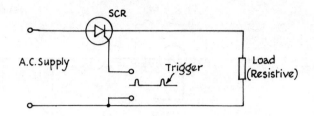

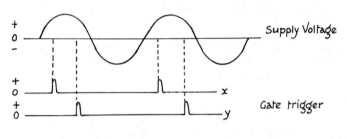

Fig. 205

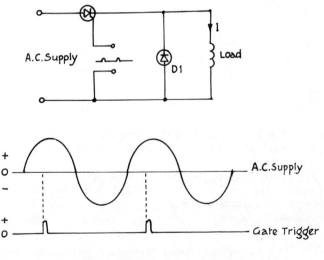

Fig. 206

154 ELECTRONICS SERVICING – QUESTIONS FOR PART 2

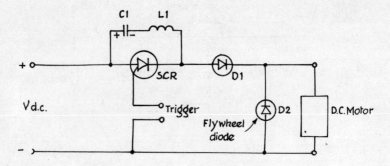

Fig. 207 Speed control of D.C. motor.

(72) Refer to Fig. 207:
    (a)   State the polarity of the trigger pulses applied.
    (b)   How is the speed of the motor altered.
    (c)   State the function of C1,L1.

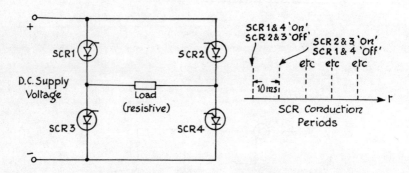

Fig. 208 Bridge invertor (commutation omitted).

(73) Refer to Fig. 208. Draw the voltage waveform across the load.
(74) Refer to Fig. 209:
    (a)   Where will the systems programme be held.
    (b)   State the function of the clock oscillator.
    (c)   Give examples of 3 commands provided by the user function buttons.
    (d)   What is the function of the R.A.M.
(75) Refer to Fig. 209:
    (a)   Give examples of 3 machine action devices.
    (b)   Which block is responsible for making decisions on the operational state of the machine.
    (c)   Give examples of 3 areas in the machine where sensors would be used.

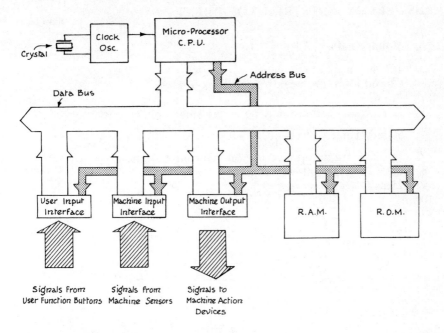

Fig. 209 Microcomputer control of working machine operation.

ANSWERS ON INDUSTRIAL EQUIPMENT

1. Voltage gain = 2.5 $\left(\dfrac{R2}{R1}\right)$.
2. 1.25V (terminal c negative w.r.t. d).
3. Closed loop, *i.e.* negative feedback applied.
4. Pin 4.
5. Zero voltage (there may be a very small voltage due to offset).
6. 500mV r.m.s. (Vo = $\dfrac{R2}{R1}$ × Vi).
7. In-phase (input applied to non-inverting terminal).
8. 1.4V (voltage-follower connection).
9. 25ms (C1×R1 secs).
10.

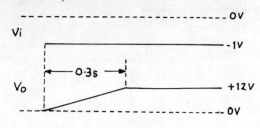

Fig. 210

11. (b)
12. 0.2V (terminal c positive w.r.t. d).
13. (a)
14. (c)
15. Ca and Ra values (external components).
16. (a) $\overline{Q}$ output high, hence TR1 conducting (Ca discharged).
    (b) Comparator 1 output low; Comparator 2 output low.
    (c) When the voltage across Ca reaches ⅔ Vcc causing comparator 1 output to go high.
17. 25 secs ($1.1 \times 25 \times 10^{-6} \times 910 \times 10^3$ secs).
18. TR1 'off'.
19. (a) TR1 'off'; TR2 'on'.
    (b) TR1 'on'; TR2 'off'.
    (c) TR2 collector.
20.

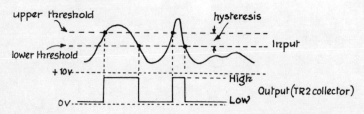

Fig. 211

INDUSTRIAL EQUIPMENT

21   With output high, input of Schmitt trigger is low and C is uncharged; C now charges via R towards output high potential. When upper threshold is reached, output goes low, C (charged) now discharges via R towards low. When lower threshold is reached, output goes high; end of cycle which is then repeated.

22

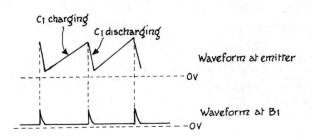

Fig. 212

23   Maximum fan-out in the low state = $\dfrac{16}{1.6} = 10$.

24
A	B	So	Co
0	0	0	0
0	1	1	0
1	0	1	0
1	1	0	1

25   Exclusive OR

26   (a) Full – adder.
   (b)
A	B	Ci	So	Co
0	0	0	0	0
0	1	0	1	0
1	0	0	1	0
1	1	0	0	1
0	0	1	1	0
0	1	1	0	1
1	0	1	0	1
1	1	1	1	1

27   (a) $S_1 = 1$
   $S_2 = 0$
   $S_3 = 1$
   $S_4 = 0$
   $C_o = 1$

   (b) Since for the column of the least significant digits there is no 'carry-in' from the previous column.

158 ELECTRONICS SERVICING – QUESTIONS FOR PART 2

28 (a) (1) E = 1   (2) F = 1   (3) D = 1
           F = 0   D = 0   E = 0
           D = 0   E = 0   F = 0
    (b) Exclusive – NOR (equality detector): $E = A.\overline{B} + B.\overline{A}$.
29 Crystal oscillator (high stability).
30 Block E is a buffer amplifier to prevent loading of the oscillator by the phase detector and Schmitt trigger.
31 (a) 5MHz
    (b)

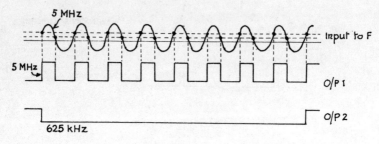

Fig. 213

32 Clocked D-type flip-flop (latch).
33
    D   Ck   Q $\overline{Q}$
    0   1   0   1
    1   1   1   0
    X   0   No change
    X = Any input
34 Shift Register (S.I.S.O.).
35

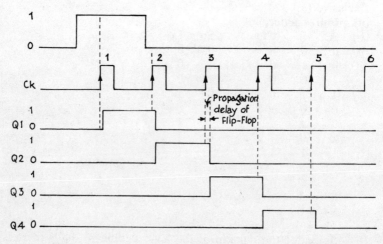

Fig. 214

INDUSTRIAL EQUIPMENT

36   (1)   Y3 = 1, all other lines = 0.
     (2)   Y7 = 1, all other lines = 0.
     (3)   Y5 = 1, all other lines = 0.
37   (1)   Terminals A, B and C permit parallel entry into the shift register.
     (2)   Terminal D allows data to be entered serially into the shift register.
38   (a)   When the enable terminal is low, the parallel inputs A, B and C are inactive and the shift register operates in the normal shift-right mode with the serial data applied to the D input; data may be read in serial form from the Qc output or in parallel form from the Qa, Qb and Qc outputs once the register is loaded. When the enable terminal is high data may be entered in parallel from A, B and C and using the PR and CLR inputs of the flip-flops.
     (b)   The PR (preset) and CLR (clear) inputs are 'forcing inputs' which allow the Q output to be either set (1) or reset (0), irrespective of its previous state.
     (c)   Suppose input A is at logic 1. When the enable terminal is set at logic 1, the preset input to latch A will be taken low and the clear input will be set high via the gates G1–G3. Thus the Q output of latch A will be set at logic 1, *i.e.* the logic state of input A. If input A is at logic 0 when the enable high is applied, the preset input will be taken to logic 1 and the clear input to logic 0. Thus the Q output will take up the logic 0 state.
39   (a)   Asynchronous (ripple-through) binary counter (3-bit).
     (b)   Logic 1 level.
     (c)   Qa = 1; Qb = 0; Qc = 1.
     (d)   The maximum counting rate is limited by the propagation delays within the counter. *i.e.* the delay between the application of a clock pulse to the counter and the counter outputs having settled.
40   (a)   Synchronous binary counter (3-bit),
     (b)

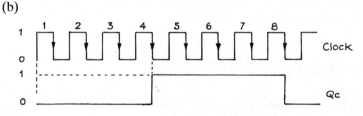

Fig. 215

     (c)   Master-slave flip-flops.
41   (a)   B.
     (b)   100.
     (c)   402.

ELECTRONICS SERVICING – QUESTIONS FOR PART 2

42 (a) 0011 0100 0101 1001.
    (b) 0001 1000 0001.
    (c) 0111 0110.

43 (a) 12.
    (b) 101.
    (c) 200.

44 (a) 255.
    (b) 58.
    (c) 301.

45 (a) 100011000.
    (b) 1111.
    (c) 1001100.

46 (a) 0100 0001 0000 0110.
    (b) 0100 1000.
    (c) 0001 0010 1001 0011.

47 (a) 27.
    (b) 42.
    (c) 56F.

48 (a) 22.
    (b) FF.
    (c) C.

49 (a)   1011
       + 1010
       10101 Answer.

    (b)   111.01
       + 1011.11
       10011.00 Answer.

    (c)   1001
       + 1101
       + 1111
       100101 Answer.

50 (a) 11100101 (minuend) – 10100011 (subtrahend)
      Form 2's complement of subtrahend:
          10100011
          01011100 (invert all bits)
             +1 (add 1 to l.s.d.)
          01011101 2's complement.

      Add 2's complement to minuend:
              11100101
           + 01011101
      Overflow 1   01000010
      (discard) Answer 01000010.

(b) 101100 − 11011
Pad minuend and subtrahend to 8 bits:
00101100 − 00011011
Form 2's complement of subtrahend:
00011011
11100100 (invert all bits)
+1 (add 1 to l.s.d.)
‾‾‾‾‾‾‾‾
11100101 2's complement
Add 2's complement to minuend:
00101100
+ 11100101
‾‾‾‾‾‾‾‾
Overflow 1   00010001

(discard) Answer 00010001.

010011 − 01110
Pad minuend and subtrahend to 8 bits:
00010011 − 00001110
Form 2's complement of subtrahend:
00001110
11110001 (invert all bits)
+1 (add 1 to l.s.d.)
‾‾‾‾‾‾‾‾
11110010 2's complement
Add 2's complement to minuend:
00010011
+ 11110010
‾‾‾‾‾‾‾‾
Overflow 1   00000101

(discard) Answer 00000101.

51 (a) (1) Thermistor; thermocouple; P–N diode.
(2) Metal Probes; thermistor (cools when immersed); float & potentiometer.
(3) Photo-transistor or diode; light-dependent resistor (L.D.R.).
(4) Light beam and photo-transistor; pressure sensitive switch; ultrasonic beam and receiver.

(b) Block B provides a reference voltage (or current) which is fed to the comparator C. If the voltage (or current) from the input transducer exceeds the reference level, then the comparator gives an output which is used to set the memory device.

52 (a) The memory of block D is used in situations where the input may return to a lower level or short duration changes are being detected, *e.g.* the interruption of a light beam. Once the input exceeds the reference level, the memory maintains the alarm in operation even though the input may fall below the reference level. A bistable or S.C.S. may be used as the memory device or 'latch'.

(b) 1kHz – 3kHz.

53 (a) G1, G2, R3, R4 and C1 form a logic gate oscillator providing a pulse output to the gate of TR1 at a frequency determined by C1, R4 values (about 1Hz).

(b) (1) With the liquid level below x-y, input A of G1 is at logic 0 and the logic gate oscillator is inactive. Thus there is no drive to TR1 which is 'off' and therefore the lamp is extinguished.

(2) When the liquid level reaches x-y, conduction takes place through the liquid causing point A to assume logic 1. This causes the logic gate oscillator to drive TR1 'on' and 'off' at 1Hz producing flashing of the indicator lamp.

(c) Power V.M.O.S. transistor (enhancement type); it has the advantage of a low current drain in the 'off' state of the alarm.

54 (a) Photo-transistor: Silicon controlled switch; light emitting diode.

(b) In the non-active state devices X, Y and Z are all 'off'. When light from the flame source falls on the base region of X, collector current and hence emitter current flows. A voltage is thereby established across R1 which is applied to the cathode gate of the S.C.S. turning it 'on'. As soon as the S.C.S. turns 'on', the circuit is complete for the forward biassed l.e.d. which emits light (red) giving a warning of the presence of a flame.

(c) When the S.C.S. turns 'on' it provides the 'latching action' for the circuit and the l.e.d. will continue to emit light even though the input flame may have been extinguished. S1 thus allows manual reset of the circuit turning the S.C.S. 'off'.

(d) When a resistor Rb is included, the circuit only responds to light levels exceeding a value set by Rb, *i.e.* Rb provides a 'threshold'. The lower the value of Rb the greater is the threshold.

55 (a) R3 is included to compensate for the temperature-sensitivity of the active strain gauge R1. The passive strain gauge is mounted close to the active strain so that it is subjected to the same temperature but is not placed under stress. Any changes in resistance of R1 and R3 due to temperature will therefore be self-cancelling in the bridge circuit.

(b) Zero voltage between points A and B and zero output voltage at point C.

56 (a) Operational amplifier connected as a difference amplifier; the main advantage is its low d.c. drift characteristic. *i.e.* the d.c. output changes very little with variations in temperature and supply voltage thus stabilising the instrument readings. Also common-mode noise or interference has little effect.

(b) P1 is used to balance out the input off-set voltage of the amplifier. It is normally adjusted to give zero output voltage with the input terminals short-circuited.

INDUSTRIAL EQUIPMENT

57   (a)   Photo-coupler (opto-coupler) consisting of a light emitting diode and phototransistor with integral lens (optical fibre light pipes may be used to link the two devices over long distances).

    (b)   A photo-coupler offers complete electrical isolation between input and output circuits, *e.g.* it may be used to block mains voltages from one circuit to another. Also the coupling medium (light) is not affected by electrical noise or interference.

58   (a)   (1)   Point C-low : TR1 – 'off'.
          (2)   Point C-high : TR1 – 'on'.
          (3)   Point C-high : TR1 – 'on'.

    (b)   R1 sets the forward current in the l.e.d. when it comes 'on'.

59   (a)   The function of the operational amplifier A is to sum the inputs from P1 and P2, *i.e.* it is connected as a summing amplifier.

    (b)   As P1 is at $-2V$ and P2 at $0V$, the sum is $-2V$ and amplifier A will give out an inverted d.c. voltage proportional to the sum (E_o). This voltage is applied via P3 to the power amplifier which drives the motor. The torque developed by the motor turns the shaft and moves the load towards the desired position. This causes P2 slider to move towards the positive supply which reduces the sum input to A. There is less error voltage (E_o) developed thus less power is fed to the motor, but it will continue to accelerate moving the load and increasing the positive potential supplied from P2 slider. When the voltage from P2 is equal to $+2V$, the sum of the inputs to A will be zero. In consequence there will be zero error voltage, no power will be supplied to the motor and no torque will be developed. The load will then be in the desired position set by P1.

60   (a)   When E_o reaches zero, the motor shaft will not stop immediately but will continue to rotate under its own inertia causing the load to overshoot the desired position. Thus the output from P2 will be greater than the desired value. This will cause a net sum input to A of reversed polarity thereby generating an error voltage of opposite sign which will cause the motor to reverse. As the load moves back towards the desired position, the error voltage will reduce and when it reaches zero the load may overshoot the desired position again. Thus the load will oscillate about its correct position. This oscillation or 'hunting' may be reduced by damping the system.

    (b)   P3 acts as an attenuator for the error voltage so that it can be set to a level to provide stable operation of the servo. If the error voltage fed to the power amplifier is too large the system may oscillate; if too small the motor may not develop sufficient torque to move the load.

61   (a)
62   (b)

63 (a) Variable-reluctance (variable-inductance) displacement transducer.
    (b) As the distance (d) is increased, the reluctance of the magnetic circuit is increased thus the inductance of the winding is reduced. Therefore with a given input voltage frequency the reactance of the winding will be less and the current reading increased.
    (c) Assuming zero resistance, the voltage will lead the current by 90°, *i.e.* voltage and current are in phase quadrature.

64 (a) Variable-permeance transducer (displacement).
    (b) As the slug is moved further into the coil its inductance is increased. Since L and C form the tank circuit of an oscillator producing a sine wave output, the effect of increasing L will be to decrease the frequency of the output (displacement is converted to change in frequency).

65 (a) Tachometer.
    (b) Frequency = Revs × No. of teeth
$$= \frac{3000 \times 6}{60} \text{Hz}$$
$$= 300 \text{Hz}.$$

66 (a) The output voltage is proportional to the rotary speed (or angular velocity).
    (b) If the direction of rotation is reversed the polarity of the output voltage pulses will reverse.

67 (a) Data logging.
    (b) The function of the multiplexer is to connect the signals from the transducers into the commoned output, one at a time in a predetermined sequence in response to digitally coded address signals.

68 (a) X-time division multiplexed samples of the analogue signals from the various transducers.
Y-time division multiplexed digital representations of the transducers signals.
    (b) The function of the sample and hold circuit is to hold a sample of the analogue signal for a sufficient time to enable the A-D converter to carry out the conversion process.

69 (a) Printer (on paper); tape or disc.
    (b) RAM.
    (c) The signal conditioning is used to make the various transducer output signals of a suitable level to apply to the system. Functions carried out by this block may include amplification, attenuation, impedance matching, filtering (to remove noise and interference) and special mathematical operations (using op-amps) to correct for transducer non-linearity or providing r.m.s. conversion, squaring *etc.*

70

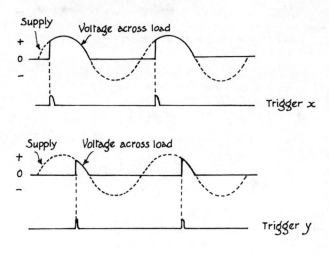

Fig. 216

71 (a) With an inductive load a back e.m.f. is produced when the S.C.R. switches 'off'. The effect of this is to maintain the flow of current through the S.C.R. for a short time after the supply voltage has become negative. The flywheel diode provides an alternative path for the flow of inductive current so that the S.C.R. may be cut-off at the commencement of the negative half-cycle of the supply voltage.

(b)

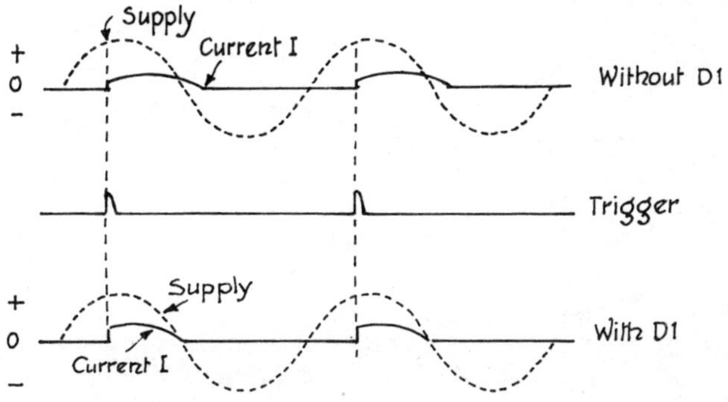

Fig. 217

72 (a) Negative (anode gate).
(b) By altering the p.r.f. of the trigger pulses to the gate; increasing the p.r.f. will increase motor speed.
(c) C1, L1 are used to provide commutation for the S.C.R., *i.e.* a means of turning 'off' the S.C.R. The values of these components determine the 'on' time of the S.C.R.

73

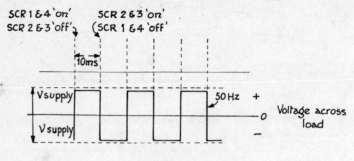

Fig. 218

74 (a) In ROM (firmware).
    (b) The function of the clock oscillator is to provide highly stable timing pulses for the systems so that all operations are carried out precisely in the correct sequence.
    (c) Half/full load; open door; wash programme (A–E); half/full spin; switch-on.
    (d) The RAM acts as a temporary store for the user commands.

75 (a) Drum motor; water pump motor; heater; relays for water valves.
    (b) C.P.U.
    (c) Water temperature (in drum); drum motor rotation; door safety mechanism.